Exogenous Enzymes as Feed Additives in Ruminants

Abdelfattah Zeidan Mohamed Salem
Abubeker Hassen • Uchenna Y. Anele
Editors

Exogenous Enzymes as Feed Additives in Ruminants

Springer

Editors
Abdelfattah Zeidan Mohamed Salem
Facultad de Medicina Veterinaria y Zootecnia
Universidad Autónoma del Estado de México
Estado de México, Mexico

Abubeker Hassen
Department of Animal Science
University of Pretoria
Pretoria, South Africa

Uchenna Y. Anele
Department of Animal Sciences
North Carolina Agricultural and Technical State University
Greensboro, NC, USA

ISBN 978-3-031-27992-8 ISBN 978-3-031-27993-5 (eBook)
https://doi.org/10.1007/978-3-031-27993-5

This Springer imprint is published by the registered company Springer Nature Switzerland AG
The registered company address is: Gewerbestrasse 11, 6330 Cham, Switzerland

Contents

Yeast as a Source of Exogenous Enzymes in Ruminant Feeding

Abdelfattah Zeidan Mohamed Salem, Moyosore Joseph Adegbeye, Mona Mohamed Mohamed Yasseen Elghandour, José Luis Ponce-Covarrubias, Andrés Gilberto Limas Martinez, Pedro Enrique Hernández Ruiz, and Deli Nazmín Tirado-González

Abstract Worldwide, the livestock industry faces various challenges, which often differ greatly depending on the geographical regions and production systems. Therefore, research and information sharing are needed for innovations and/or implementation of specific solutions to these challenges. These challenges includes: poor fibre digestibility, acidosis and endotoxins production in high milk producing cows, poor animal performance due to mycotoxin ingestion, irregular growth pattern in the dry season in some tropical countries, greenhouse gases production (nitrous oxide and methane), gastrointestinal microbial instability in weaned animals and alternatives solutions due to the ban on the use of antibiotics at subtherapeutic levels. This paper presents convincing evidence that yeast is a potential candidate to address these challenges. For instances, *Saccharomyces cerevisiae* can improve fibre digestibility and red rice yeast (*Monascus purpureus*) can reduce methane output. Similarly, *Saccharomyces cerevisiae* can protect consumers and the animal from the effect of aflatoxin by reducing its (aflatoxin) concentration in milk and increasing its concentration in faecal output. It could also be used in feed cost modulation by replacing soybean without affecting health and productivity of

A. Z. M. Salem (✉) · M. M. M. Y. Elghandour
Facultad de Medicina Veterinaria y Zootecnia, Universidad Autónoma del Estado de México, Estado de México, Mexico
e-mail: salem@uaemex.mx

M. J. Adegbeye
Department of Animal Production and Health, Federal University of Technology, Akure, Nigeria

J. L. Ponce-Covarrubias · P. E. H. Ruiz
Escuela Superior de Medicina Veterinaria y Zootecnia No. 3, Universidad Autónoma de Guerrero, Tecpan de Galeana, Guerrero, Mexico

A. G. L. Martinez
Facultad de Ingeniería y Ciencias, Universidad Autónoma de Tamaulipas, Centro Universitario "Adolfo López Mateos" Cd Victoria Tamaulipas, Mexico

D. N. Tirado-González
Departamento de Ingenierías, Tecnológico Nacional de México (TecNM)/Instituto Tecnológico El Llano Aguascalientes, Aguascalientes, Mexico

A. Z. M. Salem et al. (eds.), *Exogenous Enzymes as Feed Additives in Ruminants*, https://doi.org/10.1007/978-3-031-27993-5_1

animals. *Kluyveromyces lactiscan* could be used to improve the protein quality of crop residues by pre-treating low protein crop residues. Yeast could also be an additional source of energy for ruminants through the inclusion of oleaginous yeasts, *Cystobasidium oligophagum* and *Cryptococcus* sp. as they can produce lipids from crop residues and by-products and increase the quantity of unsaturated fatty acid escaping biohydrogenation in the rumen of ruminant. Addition of yeasts in ruminant feeding can have multifaceted benefits to the improvement of their productivity.

1 Introduction

The livestock industry has been challenged with various constraints that not only limit productivity but also has increased its role as a major source of greenhouse gases. Livestock farming worldwide is responsible for 12–18% of the anthropogenic greenhouse gas emissions depending upon the method of estimation (Westhoek et al. 2011). Possible suggestions to reduce emissions are increasing the level of productivity without increasing animal populations, improving animal output to reduce methane production per unit of product or practising climate-smart animal production, where animals are only given the nutrients they need based on precise nutrients calculation during feed formulation. Additionally, better utilisation and optimisation of the uses of crop residues and agro-industrial by-products can help increase feed biomass while reducing environmental impacts of livestock production (Salihu et al. 2012).

The ever-increasing global human population has continually increased the demand for protein, especially from meat and milk, which in turn places a huge demand for high-quality livestock feeds. Agricultural and agro-industrial residues might be potential source of high-quality livestock feeds if the nutrients embedded in the fibrous materials could be made available. Yeast may be a useful component due to its enzymatic activities and potential to convert agricultural and agro-industrial waste into higher-quality feeds. Thus, efficient recycling of agricultural waste for livestock production instead of composting the waste in the open air (thus contributing to greenhouse gas emission) could be a step forward in implementing climate-smart agriculture.

Consumers' awareness has shifted from just the consumption of protein from animal origin to genuine concern about how these products (meat and milk) are produced. Due to the link between animal protein consumption and health concerns, it is important to gain a better understanding of how levels of potentially deleterious and beneficial constituents (e.g., saturated and unsaturated fatty acids and cholesterol) in livestock protein products can be controlled by animal feeding. Furthermore, due to recent global awareness over the use of synthetic antibiotics as feed additives and potential effects of their residues on human health and water bodies, there has been a ban on the use of medically important antibiotics in the European Union and United States. Since this ban, antibiotica alternatives such as yeast, filamentous fungi, enzymes, essential oils, plant seeds, extracts, have been used as feed additives to manipulate rumen fermentation.

Another challenge is the health of animals at weaning, which occurs during transition from liquid feed to solid feed, when the digestive system changes from pre-ruminant to ruminant. The young animals are confronted with disruption in the microbial stability of the rumen which could lead to diarrhoea. The reason for such microbial imbalance when diet is changed from milk to solid might be due to the reduction in the lactic acid-producing bacteria which make the gut contents more acidic; such acidity would create an ecosystem unfavourable for growth or proliferation of pathogenic microbes (Dowarah et al. 2017). Thus, the animal system self-protects itself before it develops full immunity. Ruminants in feedlots and highly productive dairy animals also face the challenge of depressed pH in the rumen because of rapid fermentation of starch (mainly from grain); this can lead to acidosis and an increase in production of lipopolysaccharide or endotoxins due to lysis of gram-negative bacteria (Khafipour et al. 2009). Acidosis and endotoxins can affect the animal physically (lameness), the producer economically (low production) and overall health of the animal through gradual systemic failure.

In low and middle-income countries, especially across the tropics, ruminant production systems have been limited by harsh environment, poor animal genetics, low feed and forage quality, often in limited supply (Steinfeld et al. 2006). The cumulative and chronic adverse effects of these challenges can result in poor productivity and high greenhouse gas (methane, nitrous oxide, CO_2) emissions intensity from livestock operations. Throughout the world, production may be limited by low fibre digestibility in ruminant diets. This may be due to the complex blend of hemicellulose and cellulose sandwiched between lignin which limits the ability of the rumen fungi and cellulolytic bacteria to cleave the lignin bonds to access the fermentable carbohydrates.

Several dietary interventions have been investigated to improve animal productivity while reducing climate impact, and some of these interventions includes: the use of ionophores, enzymes, plant seeds, plant products and yeast. For yeast probiotics and prebiotics are perhaps very likely candidates as alternatives to synthetic antimicrobial compounds. Yeasts could be a good probiotic to use in climate-smart livestock production systems and are generally recognized as being safe. They can be useful in most production systems ranging from low to highly productive ruminants whose gut microbial stability are susceptible to disruption by high grain (*i.e.* starch) inputs (Chaucheyras-Durand et al. 2008). Nutritionally, yeasts provide protein of good quality (with high levels of threonine, lysine and methionine) and are rich in B-complex vitamins (pantothenic acid, niacin, folic acid, and biotin) and essential trace minerals. Yeasts are also capable of producing extracellular enzymes (amylases, β-galactosidase and phytases) (Barbalho 2005; Paryad and Mahmoudi 2008).

2 Composition of Yeast Culture

Yeast cell walls contain glucans, mannoproteins and chitin (Klis 1994; Moukadiri et al. 1997) and are reported to contain biologically valuable proteins, vitamin B-complex and important trace minerals that are known to promote bacteria gorwth in the rumen and production extracellular enzymes (Vohra and Satyanarayana 2001; Paryad and Mahmoudi 2008), such as lipases, amylases, β-galactosidase, xylanase and pectinase. The red yeast rice contains 2 mg/g of lovastatin, mevastatin (Miller and Wolin 2001; Wang et al. 2016) which are probably important due to its ability to reduce methane emission in ruminant. Moreover, yeast culture created in the yeast fermentation process provides a mixture of micronutrients that stimulate bacterial growth in the rumen (Callaway and Martin 1997).

3 Yeasts as a Source of Exogenous Enzymes

Microorganisms are generally responsible for degradation of plants in ruminant (Russell 2002). The ability of microbes to degrade these materials is inherently embedded in their system. Various organisms produce enzymes such as lipase, amylase, cellulase, pectinase, β-glucanase to degrade organic matter in nature. Fungi and bacteria are mainly responsible for this. Taking advantage of the ability of microbes to produce extracellular enzymes might play an important role in extracting nutrients from crop residues and by-products.

In the fungus kingdom, yeast is one of the members that are of interest because of its multifaceted functions. Over 1500 different species of yeast are known and they are widely distributed in nature. Some yeasts reside in the gut of animals (including humans) and other ecosystems where they co-habit with other microbe types. Yeast used in animal feeds could either contain live or inactivated cells. Yeasts other than *Saccharomyces cerevisiae* having probiotic properties include those belonging to the genera *Pichia*, *Metschnikowia*, *Isaatchenkia*, *Yarrowia*, *Debaryomyces*, *Candida* and *Kluyveromyces* (Vohra et al. 2016). Yeasts can produce different enzymes such as lipase, amylase, cellulase, pectinase and xylanase during fermentation processes. Other enzymes produced by yeasts include β-glucanase, esterase, β-glucosidases and protease (Maturano et al. 2012). *S. cerevisiae* IC67 and *Torulaspora delbrueckii* IC103 strains are capable of fermenting cocoa (Visintin et al. 2017). The ability of yeasts to produce lipids from plant biomass may be attributed to the inherent lipase-producing activity of the yeast which allows them to convert substrate to microbial lipids. Yeasts such as *Yarrowia lipolytica*, *Lipomyces starkeyi*, *Rhodotorula glutinis*, which can accrue intracellular lipids above 20% of their dry cell weight, have been referred to as oleaginous yeasts (Ratledge 2004) - (Table 1).

Hostinová (2002) investigated the starch-degrading activity of two strains of the yeast, *Saccharomycopsis fibuligera*. *S. fibuligera* KZ synthesized an amylolytic

Table 1 Some important enzymes of yeast origin

Enzymes	Source	References
Invertase	*Saccharomyces* sp.	Chaplin and Bucke (1990)
Lactase	*Kluyveromyces* sp.	
Lipase	*Candida* sp.	
Raffinose	*Saccharomyces* sp.	
Polygalacturonase	*Saccharomyces cerevisiae*	Poondla et al. (2016)
Xylase	*Scheffersomyces stipitis* (*Pichia stipitis*)	Jin et al. (2013)
Lipase	*Candida rugosa*	Vakhlu and Kour (2006)
Lipase	*Trichosporon asahii*	Kumar and Gupta (2008)
Lipase	*Pseudozyma hubeiensis* HB85A *Debaryomyces occidentalis*-like HB83 *Cryptococcus* sp. HB80 *Pseudozyma* sp. strains HB27B *Filobasidium floriforme*-like strain HB26B *Cryptococcus laurentii* strain HB18 *Aureobasidium* sp. strains HB07 *Cryptococcus* sp. strain HB80 *Debaryomyces occidentalis-like* strain HB83, *Candida antarctica, Candida rugosa*	Bussamara et al. (2010) Vakhlu and Kour (2006) Bussamara et al. (2010)
Lipase	*Yarrowia lipolytica, Lipomyces starkeyi* *Rhodotorulaglutinis*	Dominguez et al. (2003) Ratledge (2004)
Amylase	*Saccharomycopsis fibuligera*	Kato et al. (1976)
Xylanase	*Candida*sp., *Geotrichum* sp. *Sporopachydermis* sp. *Trichosporon* sp., *Pichia* sp. *Sugiyamaella* sp.	Wanlapa et al. (2013)
Pectinase (major among tropical yeasts)	*Kluyveromyces wickerhamii* *Kluyveromyces marxianus*	Alimardani-Theuil et al. (2011)
Xylanase, β-mannanases, laccases, cellulases, endoglucanase, β-glucosidases	*Pichia pastoris*	Ergün and Çalik (2016)
Lipase	*Candida guilliermondii*	Oliveira et al. (2014)
Glucoamylases	*Saccharomycopsis fibuligera* IFO 0111	Hostinová (2002)
α-Glucosidase	*Saccharomycopsis fibuligera* KZ	Hostinová (2002)
Cellulase	*Kluyveromyces lactis*	Jäger et al. (2011)

(continued)

Table 1 (continued)

Enzymes	Source	References
Polygalacturonase (mainly endo- polygalacturonase)	*Kluyveromyces marxianus* *Saccharomyces cerevisiae*, *Candida* sp.	Blanco et al. (1999)
Polygalacturonase, pectin esterase	*Rhodotorula* sp.	Blanco et al. (1999)
Polygalacturonase, pectin esterase, pectin lyase	*Cryptococcus cylindricus* *Cystofilobasidium lari-marini* *Cystofilobasidiumcapitatum*	Birgisson et al. (2003)
β-Glucosidases, pectinases, proteases	*Saccharomyces cerevisiae* BSc562 *Debaryomyces vanrijiae* BDv566 *Candidasake* BCs403	Maturano et al. (2015)
Xylanases, amylases	*Debaryomyces vanrijiae* BDv566 *Candida's sake* BCs403	Maturano et al. (2015)
Polygalacturonase	*Saccharomyces cerevisiae* var. *chevalieri* *Candida rugopelliculosa* *Kluyveromyces marxianus* *Kluyveromyces thermotolerans*	Alimardani-Theuil et al. (2011)
Pectinase	*Saccharomycesfragilis*	Blanco et al. (1999)
Polygalacturonase	*Cystofilobasidium infirmominiatum* *Cryptococcus adeliensis* *Guehomyces pullulans*	Cavello et al. (2017)

complex composed of α-amylase, glucoamylase and α-glucosidase, whereas *S. fibuligera* IFO 0111 produced glucoamylase.

Bussamara et al. (2010) selected 29 isolates of yeast and yeast-like strains to test, under in vitro conditions, their ability to produce lipases in the presence of bovine fat or soy oil as enzyme inducers. The most prolific lipase producers on bovine fat were *Cryptococcus* sp_1 strain HB80 (292.67 U/l), black fermenter yeast strain HB49 (298.75 U/l), *Debaryomyces occidentalis*-like strain HB83 (305.85 U/l) and *Pseudozyma hubeiensis* strain HB85A (610.40 U/l). On the soy oil, the yeasts with the highest lipase activities were *Pseudozyma* spp. strains HB27B (328.53 U/l), *Cryptococcus* sp_1 strain HB80 (350.5 U/l), *Debaryomyces occidentalis*-like strain HB83 (352.2 U/l) and *Pseudozyma hubeiensis* strain HB85A (385.6 U/l). The lipolytic activity of yeast is very important and have potential applications in the livestock industry. Triacylglycerol has a gross energy content that is 2.5 times higher than carbohydrate. Thus, it is feasible that exogenous lipase activity from yeast may help increase energy utilisation of ruminant diets, by working in synergy with bile and pancreatic lipase to hydrolyse lipids. It is also possible that dietary yeast may improve fatty acid profile available for absorption by animals. However, further research on individual yeast species and strains should be conducted to minimise any risks to animal health or safety of ruminant food products by ensuring that they do not produce any toxic metabolites.

In xylose-fermenting yeasts, the degradation of xylose substrate involves the actions of xylose reductase and xylose dehydrogenase (Sedlak and Ho 2004), each requiring the co-factors NADPH and NAD+, respectively. However, most bacteria make use of alternative pathways that involve only xylose isomerase. Both pathways result in xylulose which is further phosphorylated by the enzyme xylulose kinase; in the process, glycolytic pathway intermediates are generated (Kricka et al. 2014).

Cellulase production is common among yeast genera such as *Rhodotorula*, *Cryptococcus*, *Candida*, *Sporobolomyces* and *Aureobasidium* (Baldrian and Vendula 2008).

In studies on the potential use of yeast in biofuel production, *Kluyveromyces lactis* was shown to secrete heterologous cellulose-associated proteins at high yield, (Jäger et al. 2011). However, this yeast may have the potential to be used in pre-treatment of low protein crop residues for feeding to ruminant.

4 Production of Enzymes from Yeast

The use of chemicals in the pre-treatment of lignocellulose substrate results in the swelling of cellulose thereby increasing accessibility of hydrolytic enzymes (Chandel et al. 2012; Hong et al. 2012; Menon and Rao 2012). The release of soluble sugar from cellulose requires a wide variety of enzymes such as endoglucanase, exoglucanase and β-glucosidase (Chandel et al. 2012). Yeasts can potentially provide such enzymes. During the fermentation process in wine production, yeasts that are not of *Saccharomyces* origin produce various enzymes including β-glucosidases, proteases, pectinases, xylanases and amylases (Maturano et al. 2015).

Pectin is one of the main constituents of the cell walls and middle lamellae of higher plant cells alongside cellulose and xyloglucan (Cavello et al. 2017). Pectinolytic enzymes (pectinases) are depolymerizing enzymes that degrade pectin and are divided into depolymerizing enzymes and de-esterifying enzymes (Cavello et al. 2017). Pectinase production in yeasts is not common; only a small number of yeast species can do this (Blanco et al. 1999). The unicellular nature, simple growth rate and lack of requirement for inducers, put yeast in an advantageous position over filamentous fungi for large-scale pectinase production (Alimardani-Theuil et al. 2011). Unlike filamentous fungi, yeasts generally secrete pectin esterase (Alimardani-Theuil et al. 2011) although there are exceptions, including *Cystofilobasidium lari-marini*, *Cystofilobasidium capitatum*, *Cryptococcus cylindricus* and *Mrakia frigida* (Birgisson et al. 2003; Nakagawa et al. 2002). Depending on the environment (substrate availability) and genetics, yeasts can produce different pectolytic enzymes such as pectin esterase, polygalacturonase and pectin lyase (Blanco et al. 1999). For instance, yeasts belonging to the genera *Kluyveromyces*, *Saccharomyces* and *Candida* can produce polygalacturonase whereas *Rhodotorula* produces both polygalacturonase and pectin esterase (Blanco et al. 1999). Microbial pectinolytic enzymes play a vital role in breakdown of pectin polymers for the nutritional purposes of microbes (Poondla et al. 2016).

Cystofilobasidium infirmominiatum, *Cryptococcus adeliensis* and *Guehomyces pullulans* are examples of cold active yeasts capable of high catalytic efficiency at low and moderate temperatures (Cavello et al. 2017). They are a good source for organic matterbiodegradation and enzyme production (Birgisson et al. 2003; Margesin et al. 2005). Cold-adapted yeasts *C. infirmominiatum*, *C. adeliensis* and *G. pullulans* were tested for their ability to produce polygalacturonase at pH 5.0 and 20 °C from fruit waste (grape pomace, lemon peel, orange peel, citrus pectin and Mexican lime peel) - (Cavello et al. 2017). Results showed the ability of *G. pullulans*, *C. infirmominiatum* and *C. Adeliensis to* produce polygalacturonase from Mexican lime peel, orange peel, lemon peel, grape pomace and citrus pectin. Results also showed that *C. adeliensis* has more pectin degrading potential than *C. infirmominiatum* and *G. pullulans* in 24 h. Polygalacturonase, which is a pectinase involved in the hydrolysis of cell walls and softening of fruit during ripening by depolymerization of the middle lamella (Della Penna et al. 1990). Poondla et al. (2016) reported that *S. cerevisiae* yeast could effectively produce soluble sugar from a wide range of agricultural wastes, including seven different fruit peels, four different oilcakes, two cereal brans, tamarind kernel and apple pomace. Since the ability of *S. cerevisiae* to produce sugar from waste is considerable, there is a large potential for using yeast in ruminant feeds containing agricultural waste products in developing countries.

Lipases share about 5% of the enzyme market and, depending on their source, are classified as microbial, animal and plant lipases, primarily functioning to catalyse the hydrolysis of triglycerides (triacylglycerols, TAG) to glycerol and fatty acids (Vakhlu and Kour 2006; Salihu et al. 2012). Triacylglycerol lipases (EC 3.1.1.3) are enzymes capable of both cleaving the ester bond of triacylglycerol and catalysing ester production (Contesini et al. 2010). Oils and fats in food materials are predominantly triacylglycerols and for efficient degradation by microbes, TAG must at first be hydrolysed into glycerol and fatty acids by lipase (Matsuoka et al. 2009). Yeasts and fungi are ideal sources of extracellular lipase because they are stable, especially the endophytic yeast, *Candida guilliermondii* (Oliveira et al. 2014).

When yeast cells are initially presented with substrate, inherent lipase in the yeast hydrolyses the lipid present in the substrate at a rate capable of promoting cell growth and enzyme synthesis (Pereira-Meirelles et al. 2000). Amphiphiles produced alongside intracellular lipases produced by microbes improve lipid hydrolysis (Cirigliano and Carman 1985). Thus, with fermentation activity increasing and substrate becoming less available, synthesis of extracellular lipases becomes essential to promote uptake of substrate and continued cell survival (Salihu et al. 2012). The release of enzyme increases the chance of enzyme-substrate interaction and improves nutrient uptake (Pereira-Meirelles et al. 2000). Hence, the highest extracellular lipase activity corresponds with the late stationary growth phase (Salihu et al. 2012).

The enzymatic activities of pure cultures of *S. cerevisiae* BSc562, *Debaryomyces vanrijiae* BDv566 and *Candida sake* BCs403 to degrade grape was studied by Maturano et al. (2015). Pure cultures of *D. vanrijiae* BDv566 and *C. sake* BCs403 produced β-glucosidases, pectinases, xylanases, amylases in the ranges of

24,000–37,000, 19,000–20,000, 37,000–41,000 and 4000–5000 area under the enzyme curve (AUEC), respectively; these activity values were considerably higher than those from pure *S. cerevisiae* BSc562 or mixed cultures of the three yeasts. The high production of β-glucosidases, pectinases, proteases, xylanases and amylases by *D. vanrijiae* BDv566 and *C. sake* BCs403 suggests that these yeasts have the potential to be incorporated into ruminant diets to increase fibre degradation without lowering proteolytic activity in the rumen.

5 Yeast Enzymes and Feed Utilization

Yeasts can degrade fibre, starch and lipids through the actions of cellulases, amylases and lipases produced by the yeast cells. Sugars, starches and cellulosic materials are used in the industrial production of ethanol through fermentation processes relying on yeasts (Liu and Shen 2008). Since these substrates represent the major energy sources in the diets of farm animals, their optimal utilisation is important in the livestock industry. Digestibilities of sugar and starch can be important limiting factors for monogastrics and poultry, but in ruminants, these carbohydrates are rapidly fermented in the reticul- rumen and omasum. However, the efficiency of digestion of lignocellulose is a major concern in the ruminant livestock industry because of limitations in the ability of the rumen microbes to rapidly degrade fibre (Tables 2 and 3). Yeasts may have a role in improving the digestibility of fibre fractions in ruminat diets through the actions of cellulases, pectinases and other fibrolytic enzymes.

5.1 Dietary Fibre Utilization

In the rumen, plant cell wall polysaccharides are broken down and fermented by bacteria, fungi and protozoa (Chaucheyras-Durand et al. 2012). Cellulose and hemicellulose are components of plant cell walls which represent about 30% of most ruminant diets (Chaucheyras-Durand et al. 2008). However, insolubility and its complex structure limit plant cell wall degradation (Nagarawa et al. 1997; Forsberg et al. 2000). To optimize fibre digestion, there is a need to maintain a ruminal environment that promotes the populations of fibre-digesting microbes (Chaucheyras-Durand et al. 2012). Yeast (*S. cerevisiae*), as a directly fed microbe, is one of the probiotics that can improve the utilization of crop residues by ruminant (Adewumi 2010). Yeast-based products can help in increasing feed intake and improve digestibility when added with the diet (Jouany 2001).

In in vitro studies using rumen fluid incubations, Zain et al. (2011) incorporated yeast (*S. cerevisiae*) at levels of 0, 0.25, 0.5 and 0.75% to examine its effect on the digestibility of ammoniated rice straw. Organic matter digestibility, pH, total bacteria population of the yeast-supplement groups were higher than the control (0%

Table 2 Summary of the impact of yeast on ruminant and pre-ruminant animals (Part A)

Source	Dose (level) used	Animal type	Impacts	Reference
Saccharomyces cerevisiae	0, 5 g	West African dwarf sheep	Acid detergent fibre (ADF), Neutral detergent fibre (NDF) digestibility	Adewumi (2010)
S. cerevisiae	1×10^9, 2×10^9, 3×10^9 CFU	In vitro	ADF, NDF digestibility, improved bacteria growth	Malik and Bandla (2010)
Yeast culture *S. cerevisiae*	0, 3, 6 g	Awassi lambs	ADF, NDF digestibility, improved growth rate	Haddad and Goussous (2005)
Live *S. cerevisiae*	0, 2.5, 5 g	Friesian calves	ADF, NDF digestibility, improved feed intake, improved average daily gain, lower ammonia nitrogen	Hassan et al. (2016)
Monascus purpureus	Not specified	Goats	Reduced methane production	Wang et al. (2016)
S. cerevisiae – CNCM I–1077	1 g	Cows	Rumen pH	Doležal et al. (2011)
Active dry *S. cerevisiae* (I-1077)	8×10^9 CFU	Steers	Rumen pH, microbial proliferation	Ding et al. (2014)
Inactive dried *S. cerevisiae*	0, 33, 66, 100% soybean replacement	Saanen goats	Digestibility, similar milk performance	De Lima et al. (2012)
Powdered freeze-dried *S. cerevisiae*	0.3, 1 g/h/day	Non-lactating cows	Rumen pH, increase in volatile fatty acids (VFA)	Guedes et al. (2008)
S. cerevisiae 1077, *S. boulardii*	1 g/d	Holstein calves	Increase in VFA	Pinos-Rodriguez et al. (2008)
Live *S. cerevisiae* NCDC-49	150 g/head/day	Goats	Increased VFA, reduced NH_3-N, improved daily weight gain	Kamal et al. (2013)
Active dried *S. cerevisiae*	0, 10 g	Dairy cows	Reduced rumen lactate	Malekkhahi et al. (2016)
S. cerevisiae	0, 3 g/day YeaSacc[1026] and 0.015 g/day Levucell	Sheep	Increased organic matter fermentation and protozoa population	Angeles et al. (1998)
Live *S. cerevisiae* CBS 493.94 culture	10 g/day (5 g twice daily)	Lactating goats	Lower lactate, increased protozoa	Giger-Reverdin et al. (2004)
Live S. cerevisiae	0, 0.5, 5 g/d	Lactating Holstein cows	Increased rumen pH, increased rumen VFA production, lower D and L lactate, lower ammonia nitrogen	Pinloche et al. (2013)

Table 3 Summary of the impact of yeast on ruminant and pre-ruminant animals (Part B)

Source	Dose (level) used	Animal type	Impacts	Reference
Saccharomyces cerevisiae	1 g	Pre-weaned calf	Increased butyrate production	Xiao et al. (2016)
Edible dairy live yeast culture of *Kluyveromyces marximanus* NRRL3234 *S. cerevisiae* NCDC42 *S. uvarum* ATCC9080	1 ml/kg live weight	Lambs	Lower feed intake, increased body weight gain, improved fibre digestibility, higher rumen pH	Tripathi and Karim (2011)
K. marximanus NRRL3234 *S. cerevisiae* NCDC42 *S. uvarum* ATCC9080	1 ml/kg body weight	Weaner lambs	Increased microbial protein synthesis, *S. cerevisiae* increased average daily gain	Tripathi and Karim (2010)
S. cerevisiae	0, 10 g/d	Lactating dairy cows	Increased microbial nitrogen flow	Ouellet and Chiquette (2016)
Live *S. cerevisiae*	0, 50 g	Cows	Improved milk yield and fat	Yalçın et al. (2011)
Dead cell walls of yeast culture and that + enzymatically hydrolysed yeast	0, 56 g/h/d, 28 g/h/d	Lactating cows	Improved milk yield, fat and protein	Nocek et al. (2011)
Live yeast products	0, 3 g/cow/d	Cows	Improved milk yield, 3.5% fat corrected milk, milk protein	Rossow et al. (2018)
S. cerevisiae ITCCF 2094	5 x 10^9 Colony-Forming Unit (cfu)/h/d	Young calfs	Total body gain, feed efficiency	Panda et al. (1995)
Yeast culture (*S. cerevisiae*)	0, 30 g	Anatolian water buffalos	Improved milk yield and 4% fat corrected milk	Degirmencioglu et al. (2013)
Inactive dry yeast (*S. cerevisiae*)	0, 33, 67, 100% replacement of soybean	Lambs	Decrease in meat colour as yeast increases	Rufino et al. (2013)
S. cerevisiae 1026	0, 0.6 g/kg	Nellore steers	Lower shear force of meat, increased cooking loss	Gomes et al. (2009)
Dry yeast *S. cerevisiae*	0, 22.41 g/kg	Boer x Saanen, Saanen goat kids	Increased weight loss, tender chevon	Freitas et al. (2011)
Dried brewer's yeast *S. cerevisiae*	0.05 kg/animal/ day, increased every 10 d at the same rate	Lambs	Increased mutton dry matter, intramuscular fat, lower water holding capacity of meat	Milewski and Zaleska (2011)

yeast) diet, while the ammonia nitrogen concentrations were lower than that of the control. The mean in total bacterial population of yeast-supplemented groups was 5.37(x 10 cfu/ml) compared with 0.29 (x 10 cfu/ml) for the control group; mean ammonia nitrogen concentrations were 10.33 mM vs. 12.31 mM, for the yeast groups and the control, respectively. Organic matter, neutral detergent fibre (NDF), acid detergent fibre (ADF) and cellulose degradability in the 0.5 and 0.75% treatments were higher than in the control. Enhanced degradation of the ammoniated rice straw could mean that yeast raised the ruminal activities of cellulase, pectinase and lignase, which can cleave the bonds that hold or prevent release of nutrients. The progressive increase in NDF, ADF and cellulose degradation shows that the effectiveness of *S. cerevisiae* in digesting fibre depends strongly on its inclusion rate.

Haddad and Goussous (2005) studied the effect of yeast culture (Diamond V® YC) supplementation on digestibility of feed offered to Awassi lambs at dietary rates of 0, 3 and 6 g/d and reported that diet digestibility for both yeast culture treatments was higher than that of the control treatment. The 3 g/d dose rate improved digestibility of dry matter (DM), organic matter (OM), apparent crude protein (CP), NDF, ADF by 6.96, 5.89, 10.87, 7.68, 7.88%, respectively. Similarly, the 6 g/d yeast culture treatment increased DM, OM, apparent CP, NDF, ADF digestibilities by 5.85, 4.65, 10.02, 6.91,10.29%, respectively.

5.2 Ruminal Fermentation Characteristics

Yeasts are active agents which can have a positive effect on ruminal fermentation (Maamouri et al. 2014), stimulating the growth of beneficial rumen microbes (Denev et al. 2007). For industrial microbial fermentation purposes, including biofuel production, pretreatment of crop residues containing hemicellulose and cellulose should ideally result in enzymatic hydrolysis into simple sugars and converted by microorganisms and to various products such as amino acids, organic acids and fatty acids (Hasunuma et al. 2013). Due to the significant role of gut microbial populations in livestock feed utilisation, many studies have investigated the prospect of using yeasts to manipulate rumen fermentation to enhance animal productivity.

Ding et al. (2014) reported that the application of active dry *S. cerevisiae* CNCM I-1077 at 8×10^9 cfu/d through a cannula could modify rumen fermentation in steers. Dosing with active dry yeast increased total bacteria, total fungi and total protozoa numbers in the rumen by 28.11, 25.37 and 41.54%, respectively, when fed to animals with different ratios of concentrate and alfalfa forage. Higher level of protozoa can be an advantage in ruminant nutrition. Although protozoa may have negative effects due to their consumption of bacteria and association with archaeal and bacterial methanogens. However, protozoa can be beneficial to the rumen health in animals consuming high starch/grain diets due to their ability to engulf starch in the rumen. An increase in protozoal population may help to stabilize the rumen environment. The reason for the increase of protozoa in the yeast-treated groups may be due to the high availability of soluble sugar, the degradation of which, being enhanced by yeast. Furthermore, Preston and Leng (1987) indicated that in a fibrous diet where soluble sugar is low, protozoa population is usually less than 100,000/ml while in a diet where starch is high, they may be up to 4,000,000/ml of rumen liquor. Perhaps, yeast modifies rumen pH such that bacterial growth is increased. Due to increased release of soluble sugar during fermentation, protozoa (holotrich and entodinia) would be attracted to the sugar from where they were on the rumen wall. This reason is supported by Abe et al. (1981), who found that holotrichs were clustered on the reticulum walls of cattle fasted for one day. The increase in protozoa may be beneficial only to ruminant consuming large quantities of grain and have abundance of readily degradable sugar in their diet helping them to stabilize rumen pH so that fermenting microbes can grow rapidly. However, excessive increases in the populations of protozoa in ruminants consuming low quality diets, such as diets low in grain or readily fermentable starch, could lead to negative effects on animal productivity, because the protozoa might consume the little soluble sugar that could have been used as substrate for proliferation of rumen bacteria. Furthermore, protozoa are known to be predatory on rumen bacteria which might result in lower production of volatile fatty acids (VFA) and microbial protein. In this case, an increase in rumen protozoa may not be beneficial and efforts to reduce the protozoal population may be useful. The use of feed additives (synthetics or plant extract) to reduce rumen protozoal numbers, has sometimes resulted in improved performance (growth, milk production and feed conversion ratio).

However, during in vitro studies, rumen protozoa have been shown to be more effective than bacteria in enhancing the degradation of mycotoxins (DAS, T-2 toxin, ochratoxin A and zearalenone) (Kiessling et al. 1984). Thus, it is possible that feeding yeast to ruminant livestock may increase mycotoxin degradation by increasing protozoal numbers in the rumen.

The rumen ecosystem is constantly changing, being always in an intricate relationship with rumen pH, diet and rumen microbes. Finding a balance between these complex relationships is essential for rumen health. Rumen pH and populations of individual microbe types in the rumen ecosystem is very much dependant on the diet. Typically, high concentrate diets and high grain diets usually lead to a rapid drop in pH with potential to induce acidosis resulting in a metabolic disorder. Subacute ruminal acidosis (SARA) accounts for substantial economic losses in the dairy industry due to decreased feed intake, reduced milk production, milk fat depression, diarrhoea and laminitis (Owens et al. 1998). The SARA and acute rumen acidosis are well-known metabolic disorders in grain-fed ruminants and are prevalent in high-producing ruminants (Lettat et al. 2012). They occur when the rumen pH of a dairy cow falls below 5.6 for more than 3 hours (Gozho et al. 2005) and usually animals try to counter this by altering their feeding pattern. The changes in feeding pattern when pH drops do not only reduce feed intake, but also reduce the efficiency of the rumen fermentation due to the variation of the nutrient supply, resulting in low production and further economic losses (Cardo 2015). Beside the challenge of poor feed intake, SARA also increases the concentration of lipopolysaccharide in the gut and subsequently, in the blood, due to the lysing of gram-negative microbes, which put pressure on the liver for detoxification (Biomin 2016).

Xiao et al. (2016) studied the effect of *S. cerevisiae* yeast products on rumen fermentation of calves given milk replacer and starter diets. Calves received either 1 g/d SmartCare yeast product in the milk plus 0.5% XPC yeast product in the starter (SC1), or 1 g/d SmartCare in milk plus 1% XPC in the starter (SC2); calves on a control treatment received no yeast products. Yeast had no effect on the rumen pH, ammonia nitrogen and blood beta-hydroxybutyric acid levels. Except for rumen butyrate, dietary yeast incorporation had no effect on ruminal concentrations of individual and total VFA. The two yeast groups had a higher ($p < 0.05$) average butyrate concentration (9.39 mmol/l) than the control group (4.07 mmol/l). The increase in butyrate may assist the calves to quickly develop and renew their rumen epithelium because butyrate enhances rumen epithelial cell proliferation and papillae development. This may prepare the animal for a straight forward transition from a pre-ruminant to a ruminant. Sakata and Tamate (1978) noted that butyrate plays a key role in the postnatal development of the ruminal epithelium. Early development of rumen epithelium would help increase rapid absorption of other VFAs from the rumen due to concomitant development of the papillae, which is one of the processes through which the development of rumen function takes place. Therefore, rapid rumen development might improve early postnatal performance such as increased weight gain, feed intake and average daily gain.

Wang et al. (2016) studied the effect of red yeast rice (*Monascus purpureus*) on goats. Results indicated that red rice yeast is effective in reducing methane production during digestion. Daily methane emission was lower ($p < 0.001$) in goats fed red yeast rice (19.9 g)) compared to control goats (21.9 g)); yeast supplementation reduced CH_4 per unit DM intake by 11%. The ability of red yeast rice to reduce methanogenesis suggests that this material would be an effective feed component to reduce energy loss in ruminants, diverting such energy for economic productive purposes into meat and milk (Santra and Karim 2003).

5.3 *Microbial Protein Synthesis*

Conversion of peptides and amino acids to ammonia by rumen microbes represents an energy cost to the host animal, due to energy required for microbial protein synthesis (Wallace et al. 1997). Feeding large amounts of nitrogen can result in a high degree of excretion in urine as urea and in faeces; in manure the N is transformed via ammonia to nitrous oxide, which has 12 times the global warming potency than methane (Steinfeld et al. 2006). In grazing ruminants, about 30% of consumed N is used for production while 60% is lost (Whitehead 1995). Due to the global warming potency of the products of excreted nitrogen, strategies to reduce nitrogen loss in ruminant excreta are of paramount interest. Animals receiving yeast have been shown to have low ruminal ammonia concentrations and increasing populations of rumen bacteria which has been associated with shifts in the amino acid flow out of the rumen (Denev et al. 2007), resulting in increased flow of bacteria nitrogen to the small intestine (Erasmus et al. 1992). In an experiment using lactating rumen-fistulated dairy cows, Ouellet and Chiquette (2016) used live yeast of *S. cerevisiae* at a rate of 0 and 10 g/d in a diet containing moderately deficient metabolizable protein supply (MDSC) and a diet with highly deficient metabolizable protein supply (HDSC). Results showed a reduction in microbial nitrogen flow (308 vs. 328 g/d) in the MDSC yeast group; in contrast, yeast increased microbial nitrogen flow in HDSC by 2.45%. The increase in microbial flow in highly deficient protein diet might be attributed to the improved ammonia-nitrogen availability in the rumen. Obviously, there was a reduction in NH_3-N (highly deficient group) in the rumen of yeast supplemented diet 2.17 (mg/dl) vs. without yeast 2.85 (mg/dl), yet microbial flow increased. The reduction may be attributed to more effective use of the nitrogen by the microbes for proliferation. This suggests that at low dietary protein levels, yeasts have the capacity to improve nitrogen availability in the rumen and increase microbial flow out of the rumen. Lower NH_3-N in yeast-fed cows suggests that yeast was able to reduce NH_3-N by improved synergy with the gut microbial community which caused their increase because of nutrient availability through increased feeding. The increased microbial proliferation might also add to microbial protein which would increase amino acids reaching the small intestine for absorption.

5.4 *Milk Production, Composition and Quality*

Dietary supplementation with *S. cerevisiae* fermentation products increases milk yield in dairy cows (Poppy et al. 2012), which could be due to the stabilizing effect of yeast products on rumen conditions (Williams et al. 1991) through enhancement in the numbers of lactate-utilizing microbes and growth of fibre digesting bacteria (Callaway and Martin 1997). Additionally, another benefit of feeding live yeast is through improved growth of *Butyrivibrio* spp. in the rumen, which facilitates the conversion of linoleic acid to conjugated linoleic acid (CLA) (Chaucheyras-Durand et al. 2008); CLA in ruminant food products has been beneficial to human health.

Yalçın et al. (2011) used a commercial live yeast product from *S. cerevisiae* (RumiSacc) that consisted of live yeast, autolyzed yeast and modified dried vinasse RumiSacc (50 g/cow) which was applied by top-dressing method to the feed and improved milk yield by 6.3% ($p < 0.05$). However, it had no effect on DM intake, feed conversion ratio (kg milk/kg DM intake) or milk composition, except for increased levels of α-linolenic acid (18:3, n-3). The increase in milk yield suggests an increase in feed digestibility and fibre digestibility because the yeast supplement did not affect feed intake. Thus, feed efficiency was enhanced. Yeast has the potential to improve milk and meat quality by altering the fatty acid profile of these products, and through the possibility of enhanced production of CLA, especially the cis9, trans11 isomer (Chaucheyras-Durand et al. 2008). Doreau and Ferlay (1994) stated that there is 95% disappearance of α-linolenic acid 18:3 (n-3) in the rumen due to biohydrogenation. The increase in the level of 18:3 (n-3) in milk (Yalçın et al.

2011) suggested that yeast could reduce the degree of biohydrogenation of this fatty acid; however, yeast supplementation had no effect on the levels of any other fatty acids (short, medium and long chain) in milk.

In another study, Nocek et al. (2011) fed multiparous lactating cows with dead cell walls of yeast culture (YC) at 56 g/cow/day and dead cell walls of yeast culture + enzymatically hydrolyzed yeast (YC + EHY) at 28 g/cow/day. Compared with the control treatment, YC and YC + EHY increased milk yield by 3.46 and 3.95%, fat-corrected milk yield by 3.85 and 4.33%, energy-corrected milk yield by 4.16 and 4.65%, respectively. Furthermore, YC and YC + EHY raised milk fat yield by 4.73 and 5.41% and milk protein yield by 3.42 and 5.98%, respectively, compared to those of the control. However, yeast product supplementation had no effect on milk fat content, although treatment YC + EHY increased milk protein levels by 2.41%. The increases in milk yield resulting from yeast supplementation may perhaps be attributed to improved fibre digestibility and enhancement in the supply of protein and amino acid through microbial protein synthesis, in terms of both quantity and quality. There is a well-known relationship between content of NDF in diet and milk fat concentration in dairy cow (Emery 1988) and sheep (Nudda et al. 2004), thus, a better NDF digestibility would lead to increased milk fat.

5.5 *Growth Performance*

In the developing world, except for periods in the year when forages are growing afresh during rainfall after a short period of drought, ruminants usually graze on low quality forages (low in protein and high in fibre). The situation becomes worse during the dry season, when they may be given crop residues (groundnut haulms, cowpea husk, sorghum husk, straws, rice bran, wheat straw, etc.) that have a high fibre content, which may result in poor growth or weight reduction in animals. Improving feed efficiency could help reduce the cost of production per unit of food from animals.

Adewumi (2010) showed that dietary yeast supplementation affected the average daily live weight gain of West African Dwarf sheep. The average daily gain was decreased by 10.86% in sheep fed groundnut haulm, increased by 13.8% in those fed guinea corn waste and 124.82% in those fed cowpea husk, compared to the group without yeast supplementation. The reason for the decrease in sheep fed groundnut haulm could be that the yeast improved feed digestibility, but caused the release of one or more anti-nutrients, such as alkaloids and saponins, which are present in many crop by-products; these metabolites can bind feed nutrients, making them inaccessible to animals. This may also explain the observation that N retention in the group without yeast feeding was higher than in the group fed with yeast. Thus, in plants with high levels of anti-nutritional factors, there might be a need to pre-treat them with yeast to reduce their anti-nutrient content before feeding them to animals. (Hasunuma et al. 2013).

It is normal for *Lactobacilli* to colonise the gut of livestock from milk feeding; this may help prevent the growth of invading pathogenic microbes. Lactose in milk favours the growth of lactic acid producers and utilizers which help to reduce pH in the gut, thus preventing the excessive proliferation of opportunistic pathogens. However, changes occur in the gut microbes when animals are transformed from liquid to solid diets due to the alteration in the microbial populations associated with different state of diet (Chaucheyras-Durand et al. 2008). It is important that during weaning, that there is a gradual shift in balance of gut microbes; animals usually react to sudden intestinal microbial disruption by diarrhoea or refusal to eat which affects their overall growth and/or performance (Chaucheyras-Durand et al. 2012). Transition from liquid (milk) to solid diets might also result in establishment of unwanted intestinal microbes resulting in poor performance (Tripathi and Karim 2010). These authors used yeast culture (*Kluyveromyces marximanus* NRRL3234 (KM), *S. cerevisiae* NCDC42 (SC), *S. uvarum* ATCC9080 (SU) and a mixture of the three yeasts in equal ratio) fed to Malpura weaned lambs. The daily dose of yeast cultures was given by drench at 1 ml/kg body weight. Finishing weight 33.2, 32.0, 30.0, 29.2 and 27.9 kg for SC, mixed yeast, SU, control and KM respectively was similar ($p > 0.05$) across treatments. Average daily gain in all yeast-supplemented groups was greater than in the control except for KM in 91 days. Feed intake in other yeast supplemented groups was lower than the control except in the SC which was the same as the control treatment. SC, SU and a mixture of the three-yeast had better daily live weight gain than the control ($p < 0.05$), while KM had the lowest live weight gain probably due to the low feed intake. Consequently, all yeast treatments had better feed efficiency than the control. This showed that the yeast supplements increased the efficiency of feed consumed by animal; this was likely due to changes in the ruminal fermentation process. In some tropical countries, for example, Nigeria, there is a dubious view of feeding high quality protein, such as soybean, to ruminants due to its high cost and documented knowledge that proteolytic rumen microbes would deaminate it, resulting in a waste of resources. Yeast has potential as a protein source in ruminant feeds. De Lima et al. (2012) used dried yeast produced from sugar cane as source of protein in comparison to soybean to feed primiparous Saanen goats. Dry yeast can contain about 42% CP (Butolo 2002), while the CP content of soybean meal can be as much as 44% (Adewumi 2010). In diets for West African dwarf sheep, dry yeast supplementation was able to increase DM, OM and total carbohydrate digestibility to a greater extent than the control group fed soybeans.

Chevon (goat meat) is the second most consumed ruminant meat after beef in Nigeria. Because it is considered to have a superior flavour, chevon is generally preferred to beef. Consumers will pay a premium to acquire chevon in local canteens and event centres across the nation. However, chevon is a more expensive product than beef. Hence improving goat nutrition to reduce production costs might encourage more farmers to go into goat production to meet the demand for goat meat. Kamal et al. (2013) used live *S. cerevisiae* (NCDC-49) as a supplement to feed local goats. The DM intake was similar between supplemented and control groups. Average daily gain of goats fed yeast (43.81 g/d) was higher than the control goats (35.05 g/d) ($p < 0.01$). Yeast improved production by reducing the feed conversion ratio of the yeast-fed group (7.65) compared to that of the control treatment group (8.89).

Degirmencioglu et al. (2013) added yeast culture (*S. cerevisiae*) to the diet of Anatolian water buffalo at 0 and 30 g yeast which accounts for 2.1×10^{10} fu/g/d of yeast. Anatolian water buffalo are dual purpose farm animals suitable for farmers living in a region with swamps and streams. Yeast supplementation improved milk yield (7.13 vs 6.22 kg/day in the control group), but had no effect on milk solid non-fat, fat, protein and lactose contents. The increase in milk yield suggested that yeast improved feed and fibre digestibility, feed efficiency, fermentation parameters and increased microbial population of the rumen.

5.6 Meat Quality

The type of feed consumed by animals can influence the quality of meat produced. Various components of animal feed can affect the properties of meat in several ways, including water-holding capacity, pH, palatability as human food, proximate composition and leanness (*i.e.* the balance between fat and muscle); some components of meat, such as CLA, may also have anticholesterolemic potential. Supplementation of animal diets with yeast may influence the chemical composition of meat through the effect of yeast on the rate of rumen fermentation and metabolism of end products (Titi et al. 2008a).

In the study of Rufino et al. (2013), inactive dry yeast (IDY - *S. cerevisiae*) was used to substitute soybean meal in lambs' diets at 0, 33, 67 and 100% replacement levels. IDY has about 35% CP (Valadares Filho et al. 2010) and its replacement had a minor effect on meat lightness score and no effect on meat yellowness and redness scores. Soft fat tissue declined with increase in IDY inclusion ($p = 0.035$), most likely because of yeast having an inferior amino acid profile (methionine and cysteine) to that of soybean meal which could have inhibited microbial proliferation and production of acetate acid in the rumen, resulting in reduced fat deposition in the carcass (Ramos et al. 2011; Rufino et al. 2013).

Titi et al. (2008b) observed that supplementation of finishing diets with yeast for dairy calves had no effect on the water-holding capacity of meat, and meat lightness and redness; however, meat yellowness of the yeast treatment groups was lower than in the control groups ($p < 0.05$). This indicates that the meat of the yeast-fed calves was lighter in colour than that of the control animals; this may have been due to lower redness or blood pigmentation, or a higher intramuscular fat content in the yeast-fed animals. In the study of Gomes et al. (2009), supplementation of a maize-based diet with a yeast culture (0.6 g/kg), given as a finishing ration to feedlot-fed Nellore steers had no effect on meat quality.

Goat meat is characterized by having protein of high biological value (Amaral et al. 2007) and a lower cholesterol content than mutton and beef (40 vs 62 vs 70 mg/100 g respectively) (Souza and Visentainer 2006). Goat meat is also lower in fat, which results in a reduction in the proportion of saturated fat and energy content when compared to other red meats (Madruga et al. 1999; Malan 2000). Freitas et al. (2011) studied the effect on meat quality of replacing soybean meal with dried yeast in the diets of goat kids. Cooking loss of meat from yeast-fed goats was 26.39% vs 27.17% in meat from control (soy bean diet) animals. The shear force of meat from the yeast-fed animals was 6.88 kg and that of meat from control animals was 5.45 kg; according to criteria described by Souza et al. (2004), the meat from both yeast-fed and control goats could be classified as tender. Yeast feeding had no effect on the moisture, protein, ash and total lipids of *Longissimus dorsi* muscle.

Milewski and Zaleska (2011) studied the effect of dietary supplementation with dried yeast on meat quality in lambs. Dry matter and intramuscular fat contents of meat from lambs fed yeast were higher but had a lower water holding capacity than meat from the control animals. Colour-space measurements, L* and a* differed between the yeast-fed and control lambs (L*: 46.01 and 48.48; a*: 12.28 and 11.14, respectively) ($p < 0.01$); yeast supplementation had no effect on colour-space parameter b*, meat cooking loss, drip loss and shear force. On sensory evaluation, dietary yeast had no effect on aroma intensity, aroma desirability, taste intensity, taste desirability, tenderness and juiciness. The lower water holding capacity of meat from yeast-fed lambs suggests that there was reduced ability of the muscle cells to retain water during heat treatment (Aaslyng et al. 2003; Micklander et al. 2005). This implies that during cooking, more minerals and other important nutrients would be lost from meat from yeast-fed lambs when cooked. Higher *a** is an indication of higher level of red colour component of the meat.

6 Source of Dietary Lipid/Microbial Lipid

Yeasts mainly accumulate lipids that comprise of saturated and unsaturated fatty acids in the form of triacylglycerol (Vyas and Chhabra 2017). Microbial lipids might be a potential alternative feedstock for energy sources, such as biodiesel (Chang et al. 2015). Microorganisms accumulating lipids more than 20% of their biomass are referred to as being oleaginous species (Ratledge and Wynn 2002). Interestingly, most of these lipids are triacylglycerols containing long-chain fatty acids comparable to common vegetable oils (Rattray 1988) which may be an alternative to the use of oil in animal feed. Addition of such yeast to diets containing crop residues or inoculating crop residues with yeast might help improve energy supply of these feed sources. For biodiesel production, Vyas and Chhabra (2017) isolated an oleaginous yeast (*Cystobasidium oligophagum* JRC1; Basionym: *Rhodotorula oligophaga*) from a soil rich in cellulosic waste and tested its ability to produce lipase and convert different carbon sources into lipids. The yeast was able to grow and accumulate lipid on a range of substrates including glucose (39.44% accumulated lipid), carboxymethylcellulose (36.46%), glycerol (42.04%), starch (41.54%), xylose (36.24%), maltose (26.31%), fructose (24.29%), lactose (21.91%) and sucrose (21.72%). The observation that fatty acid profiles of accumulated lipids were suitable for biodiesel production suggests that this oleaginous yeast would be of value in the ruminant livestock feed industry. The wide range of substrates that this yeast can use is compatible with those that are usually present in potential ruminant feed ingredients as well as those present in the rumen. Thus, the yeast can potentially be used in pre-treatment of feeds. It could serve as an alternative to the use of dietary oil, which is on demand for use in human foods. Since supplementation of ruminant diets with lipid-rich feeds, as extracted lipids, or whole oil seeds, is another strategy for modifying ruminal biohydrogenation, it is possible that feeding an oleaginous yeast would increase the lipid content in the rumen and enable more fatty acids to escape biohydrogenation; increase the supply of unsaturated fatty acid to the small intestine for absorption may result in meat or milk with a fatty acid profile beneficial to human health. The lipid profile of oleaginous yeast species can be altered by changing their substrate carbon sources (Sitepu et al. 2013). Fat supplementation of diets for dairy cattle can change the fatty acid profile of milk and improve milk yield, depending on the type of fat supplementation and concentration being ingested by a cow (Dschaak 2012; Patra 2013). The ability of yeast to change its lipid profile can be a novel way of enhancing quality and responding to changes in consumers' tastes in dairy and other ruminant products. This could be effected by using yeast as a feed supplement to modify the fatty acid profile passing to the small intestine of animals for absorption. These are convincing results, suggesting that yeast biomass containing high lipids can be obtained by pre-treating agricultural crop residue instead of feeding the residue directly to animal but more research is needed in this area. With this development, there may not be a need to use fat supplements, or the fat content of ruminant diets could be reduced, to take advantage of the ability of the yeast if they could increase fatty acid delivery.

7 Means of Counteracting the Negative Effects of Mycotoxins

Many of the non-conventional plants used as feeds contain high levels of plant secondary metabolites such as phyate, cyanogens, tannins, alkaloids, oxalates etc., which may be anti-nutritional depending on the quantity in the feed. It is possible that yeasts can degrade plant phytochemicals to both detoxify them and produce important nutrients for animals. Hence, by pre-treatment of agricultural by-products with yeast it should be possible to improve the nutritive value of these plants.

Mycotoxins are pathogenic by-products of fungal growth on a range of substrates and is often present in poorly stored feed which allowed moulds to grow; this can affect animals' health and performance. Greater production demands and changes in ruminant feeding practices have increased opportunities for mycotoxins to have negative effects on production and animal health (including kidney, liver and gastrointestinal functions) (Gott 2017). Some mycotoxins affecting ruminants include deoxynivalenol, zearalenone, aflatoxin B1, T-2 toxin and ergot alkaloids.

Aflatoxins are by-products of *Aspergillus flavus* and *Aspergillus parasiticus* metabolism (Firmin et al. 2011). When ingested by ruminants, aflatoxin B1 is partly metabolized in the liver to a more polar compound: aflatoxin M1, which is more toxic than the parent toxin and can be excreted in urine, faeces and milk (Allcroft et al. 1967). In Europe, contaminated milk and other products derived from it are the primary sources of aflatoxin consumption by humans (Prandini et al. 2009).

Firmin et al. (2011) studied the ability of modified *S. cerevisiae* cell wall to neutralize aflatoxin B1. In this study, ewes were given constant daily doses of aflatoxin B1-contaminated wheat at 60μg/kg of feed. Modified yeast, dosed together with aflatoxin B1-contaminated wheat, tended to reduce aflatoxin M1 excretion in urine, but aflatoxin M1 excretion in faeces was almost doubled ($p < 0.05$). The yeast treatment also increased the faecal excretion of aflatoxin B1 ($p < 0.05$). Yeast appeared to reduce the absorption of aflatoxin through the rumen epithelium into the blood stream, which reduced the load in liver for detoxification. The reduced absorption in the rumen led to the greater excretion of aflatoxin in faeces and thus reduced the risk of contamination of milk and meat.

8 Conclusions

Inclusion of yeast in diet improve the digestibility of neutral detergent fiber and acid detergent fiber, rumen microbial population and enhance ruminal pH stability in starch-based diet through increased population of protozoa. Pre-treatment of ruminant feed ingredient with oleaginous yeast could serve as a source lipid supply to ruminant. Inclusion of live yeast in ruminant diet could improve animal product quality by improving the fatty acid profile of meat and milk. On greenhouse gases, red yeast rice *M. purpureus* can reduce methane output when given to ruminant. Inclusion of yeast in animal diet has the potential to enhance degradation of mycotoxin and neutralization of Aflatoxin B_1 which would play a role in animal health and safety of animal products for consumer.

References

Aaslyng MD, Bejerholm C, Ertbjerg P et al (2003) Cooking loss and juiciness of pork in relation to raw meat quality and cooking procedure. Food Qual Prefer 14:277–288

Abe M, Iriki T, Tobe N et al (1981) Sequestration of holotrich protozoa in the reticulo-rumen of cattle. Appl Environ Microbiol 41:758–765

Adewumi MK (2010) Nutritive value of some crop residues with or without baker's yeast (*Saccharomyces cerevisiae*) as supplements for west African dwarf sheep. Trop Anim Prod Invest 13:7–12

Alimardani-Theuil P, Gainvors-Claisse A, Duchiron F (2011) Yeasts: an attractive source of pectinases – from gene expression to potential applications: a review. Process Biochem 46:1525–1537

Allcroft R, Roberts BA, Butler WH (1967) Aflatoxin in milk. Food Cosmet Toxicol 5:597–598

Amaral CMC, Pelicano ERL, Yáñez EA et al (2007) Caracteristicas de carcaça e qualidade de carne de cabritos Saanen alimentados com ração completa farelada, peletizada e extrusada. Ciência Rural 37:550–556

Angeles SCA, Mendoza GBM, Cobos MAP et al (1998) Comparison of two commercial yeast cultures (*Saccharomyces cerevisiae*) on ruminal fermentation and digestion in sheep fed on corn-stover diet. Small Ruminant Res 31:45–50

Baldrian P, Vendula V (2008) Degradation of cellulose by *basidiomycetous* fungi. FEMS Microbiol Rev 32:501–521

Barbalho R (2005) Levedura inativa como microingrediente de ação profilática na alimentação de aves e suínos. Guia Avicultura Indus 6:40–46

Biomin (2016) What's wrong with my herd? Part 2: Endotoxins Science and Solutions. A magazine of Biomin®: Ruminant #37

Birgisson H, Delgado O, Garcia Arroyo L et al (2003) Cold-adapted yeasts as producers of cold-active polygalacturonases. Extremophiles 7:185–193

Blanco P, Sieiro C, Villa TG (1999) Production of pectic enzymes in yeasts. FEMS Microbiol Lett 175:1–9

Bussamara R, Fuentefria AM, de Oliveira E et al (2010) Isolation of a lipase-secreting yeast for enzyme production in a pilot-plant scale batch fermentation. Bioresour Technol 101:268–275

Butolo JE (2002) Ingredientes de origem vegetal. In: Qualidade de ingredientes na alimentação animal. CBNA, Campinas, pp 93–237

Callaway TS, Martin SA (1997) Effects of a *Saccharomyces cerevisiae* culture on ruminal bacteria that utilize lactate and digest cellulose. J Dairy Sci 80:2035–2044

Cardo L (2015) Tackling acidosis the dangers of SARA. Science and solutions: a magazine of biomin®: ruminant #25

Cavello I, Albanesi A, Fratebianchi D et al (2017) Pectinolytic yeasts from cold environments: novel findings of *Guehomyces pullulans*, *Cystofilobasidium infirmominiatum* and *Cryptococcus adeliensis* producing pectinases. Extremophiles 21:319–329

Chandel AK, Chandrasekhar G, Silva MB et al (2012) The realm of cellulases in biorefinery development. Crit Rev Biotechnol 32:187–202

Chang YH, Chang KS, Lee CF et al (2015) Microbial lipid production by oleaginous yeast *Cryptococcus* sp. in the batch cultures using corn cob hydrolysate as carbon source. Biomass Bioenergy 72:95–103

Chaplin MF, Bucke C (1990) Enzyme technology. Cambridge University Press

Chaucheyras-Durand F, Walker ND, Bach A (2008) Effects of active dry yeasts on the rumen microbial ecosystem: past, present, and future. Anim Feed Sci Technol 145:5–26

Chaucheyras-Durand F, Chevaux E, Martin C et al (2012) Use of yeast probiotics in ruminants: effects and mechanisms of action on rumen pH, fibre degradation, and microbiota according to the diet. In: Probiotics in animals (In Tech), pp 119-152

Cirigliano MC, Carman GM (1985) Purification and characterization of liposan, a bioemulsifier from *Candida lipolytica*. Appl Environ Microbiol 50:846–850

Contesini FJ, Lopes DB, Macedo GA et al (2010) *Aspergillus* sp. lipase: potencial biocatalyst for industrial use. J Mol Catal B Enzym 67:163–171

De Lima LS, Alcalde CR, Freitas HS et al (2012) Performance of dairy goats fed diets with dry yeast from sugar cane as protein source. R Bras Zootec 41:232–236

Degirmencioglu T, Ozcan T, Ozbilgin S et al (2013) Effects of yeast culture addition (*Saccharomyces cerevisiae)* to Anatolian water buffalo diets on milk composition and somatic cell count. Mljekarstvo 63:42–48

Della Penna D, Lashbrook CC, Toenjes K et al (1990) Polygalacturonase isozymes and pectin depolymerization intrans- genicrin tomato fruit. Plant Physiol 94:1882–1886

Denev S, Peeva A, Radulova P et al (2007) Yeast cultures in ruminant nutrition. Bulg J Agric Sci 13:357–374

Ding G, Chang Y, Zhao L et al (2014) Effect of *Saccharomyces cerevisiae* on alfalfa nutrient degradation characteristics and rumen microbial populations of steers fed diets with different concentrate-to-forage ratios. J Anim Sci Biotech 5:24–44

Doležal P, Dvořáček J, Doležal J et al (2011) Effect of feeding yeast culture on ruminal fermentation and blood indicators of Holstein dairy cows. Acta Vet Brno 80:139–145

Dominguez A, Costas M, Longo MA et al (2003) A novel application of solid-state culture: production of lipases by *Yarrowia lipolytica*. Biotechnol Lett 25:1225–1229

Doreau M, Ferlay A (1994) Digestion and utilization of fatty acid by ruminants. Anim Feed Sci Technol 45:379–396

Dowarah R, Verma AK, Agarwal N (2017) The use of *lactobacillus* as an alternative of antibiotic growth promoters in pigs: a review. Anim Nutr 3:1–6

Dschaak CM (2012) Use of rumen modifiers to manipulate ruminal fermentation and improve nutrient utilization and lactational performance of dairy cows. PhD thesis, Utah State University

Emery RS (1988) Milk fat depression and the influence of diet on milk composition. Vet Clinic North Am. Food Anim Pract 4:289–305

Erasmus LJ, Botha PM, Kistner A (1992) Effect of yeast culture supplement on production, rumen fermentation and duodenal nitrogen flow in dairy cows. J Dairy Sci 75:3056–3065

Ergün BG, Çalik P (2016) Lignocellulose degrading extremozymes produced by *Pichia pastoris*: status and prospects. Bioprocess Biosyst Eng 39:1–36

Firmin S, Morgavi D, Yiannikouris A et al (2011) Effectiveness of modified yeast cell wall extracts to reduce aflatoxin B1 absorption in dairy ewes. J Dairy Sci 94:5611–5619

Forsberg CW, Forano E, Chesson A (2000) Microbial adherence to the plant cell wall and enzymatic hydrolysis. In: Ruminant physiology: digestion, metabolism, growth, and reproduction. CABI Publishing, Oxon, pp 79–98

Freitas HS, Alcalde CR, de Lima LS et al (2011) Quantitative characteristics of carcass and meat quality of ¾ Boer + ¼ Saanen and Saanen goat kids fed diets with dry yeast. R Bras Zootec 40:630–638

Giger-Reverdin S, Sauvant D, Tessier J et al (2004) Effect of live yeast culture supplementation on rumen fermentation in lactating dairy goats. S Afr J Anim Sci 34:59–61

Gomes RC, Leme PR, Silva SL et al (2009) Carcass quality of feedlot finished steers fed yeast, monensin, and the association of both additives. Arq Bras Med Vet Zootec 61:648–654

Gott P (2017) The performance inhibitor hiding in feed. Science and solutions: a magazine of biomin™: ruminant #49

Gozho GN, Plaizier JC, Krause DO et al (2005) Subacute ruminal acidosis induces ruminal lipopolysaccharide endotoxin release and triggers an inflammatory response. J Dairy Sci 88:1399–1403

Guedes CM, Goncalves D, Rodrigues MAM et al (2008) Effects of a *Saccharomyces cerevisiae* yeast on ruminal fermentation and fibre degradation of maize silages in cows. Anim Feed Sci Technol 145:27–40

Haddad SG, Goussous SN (2005) Effect of yeast culture supplementation on nutrient intake, digestibility, and growth performance of Awassi lambs. Anim Feed Sci Technol 118:343–348

Hassan A, Azm S, Kholif AE et al (2016) Performance of crossbred dairy Friesian calves fed two levels of *Saccharomyces cerevisiae*: intake, digestion, ruminal fermentation, blood parameters and faecal pathogenic bacteria. J Agric Sci 154:1488–1498

Hasunuma T, Okazaki F, Okai N et al (2013) A review of enzymes and microbes for lignocellulosic biorefinery and the possibility of their application to consolidated bioprocessing technology. Bioresour Technol 135:513–522

Hong F, Guo X, Zhang S et al (2012) Bacterial cellulose production from cotton-based waste textiles: enzymatic saccharification enhanced by ionic liquid pretreatment. Bioresour Technol 104:503–508

Hostinová E (2002) Amylolytic enzymes produced by the yeast *Saccharomycopsis fibuligera*. Biologia, Bratislava 57:247–251

Jäger G, Girfoglio M, Dollo F et al (2011) How recombinant swollenin from *Kluyveromyces lactis* affects cellulosic substrates and accelerates their hydrolysis. Biotechnol Biofuels 4:33
Jin M, Sarks C, Gunawan C et al (2013) Phenotypic selection of a wild *Saccharomyces cerevisiae* strain for simultaneous saccharification and co-fermentation of AFEX™ pretreated corn Stover. Biotechnol Biofuels 6:108–122
Jouany JP (2001) A new look to at yeast culture as probiotics for ruminants. Feed Mix 9:17–19
Kamal R, Dutt T, Singh M et al (2013) Effect of live *Saccharomyces cerevisiae* (NCDC-49) supplementation on growth performance and rumen fermentation pattern in local goat. J Appl Anim Res 41:285–288
Kato K, Kuswanto K, Banno I et al (1976) Identification of *Endomycopsis fibuligera* isolated from ragi in Indonesia and properties of its crystalline glucoamylase. J Ferment Technol 54:831–837
Khafipour E, Krause DO, Plaizier JC (2009) A grain-based sub-acute ruminal acidosis challenge causes translocation of lipopolysaccharide and triggers inflammation. J Dairy Sci 92:1060–1070
Kiessling K, Pettersson H, Sandholm K et al (1984) Metabolism of aflatoxin, ochratoxin, zearalenone, and three trichothecenes by intact rumen fluid, rumen protozoa, and rumen bacteria. Appl Environ Microbiol 47:1070–1073
Klis FM (1994) Review: cell wall assembly in yeast. Yeast 10:851–869
Kricka W, Fitzpatrick J, Bond U (2014) Metabolic engineering of yeasts by heterologous enzyme production for degradation of cellulose and hemicellulose from biomass: a perspective. Front Microbiol 5:1–11
Kumar SS, Gupta R (2008) An extracellular lipase from *Trichosporon asahii* MSR 54: medium optimization and enantioselective deacetylation of phenyl ethyl acetate. Process Biochem 43:1054–1060
Lettat A, Noziere P, Silberberg M et al (2012) Rumen microbial and fermentation characteristics are affected differently by bacterial probiotic supplementation during induced lactic and subacute acidosis in sheep. BMC Microbiol 12:142
Liu R, Shen F (2008) Impacts of main factors on bioethanol fermentation from stalk juice of sweet sorghum by immobilized *Saccharomyces cerevisiae* (CICC 1308). Bioresour Technol 99:847–854
Maamouri O, Selmi H, M'hamdi N (2014) Effects of yeast (*saccharomyces cerevisiae*) feed supplement on milk production and its composition in Tunisian Holstein Friesian cows. Sci Agric Bohem 45:170–174
Madruga MS, Arruda SGB, Araújo LT et al (1999) Efeitoda idade de abate no valor nutritivo y sensorial da carne caprina de animais mestiçoes. Ciênc Tecnol Alimentos 19:374–379
Malan SW (2000) The improved Boer goat. Small Ruminant Res 36:165–170
Malekkhahi M, Tahmasbi AM, Naserian AA et al (2016) Effects of supplementation of active dried yeast and malate during sub-acute ruminal acidosis on rumen fermentation, microbial population, selected blood metabolites, and milk production in dairy cows. Anim Feed Sci Technol 213:29–43
Malik R, Bandla S (2010) Effect of source and dose of probiotics and exogenous fibrolytic enzymes (EFE) on intake, feed efficiency, and growth of male buffalo (*Bubalus bubalis*) calves. Trop Anim Health Prod 42:1263–1269
Margesin R, Fauster V, Fonteyne PA (2005) Characterization of cold active pectate lyases from psychrophilic *Mrakia frigida*. Lett Appl Microbiol 40:453–459
Matsuoka H, Miura A, Hori K (2009) Symbiotic effects of a lipase-secreting bacterium, *Burkholderia arboris* SL1B1, and a glycerol-assimilating yeast, *Candida cylindracea* SL1B2, on triacylglycerol degradation. J Biosci Bioeng 107:401–408
Maturano YP, Rodrıguez Assaf LA, Toro ME et al (2012) Multi-enzyme production by pure and mixed cultures of *saccharomyces* and non-*saccharomyces* yeasts during wine fermentation. Int J Food Microbiol 155:43–50
Maturano YP, Assof M, Fabani MP et al (2015) Enzymatic activities produced by mixed *saccharomyces* and non-*saccharomyces* cultures: relationship with wine volatile composition. Antonie Van Leeuwenhoek 108:1239–1256

Menon V, Rao M (2012) Trends in bioconversion of lignocellulose: biofuels, platform chemicals & biorefinery concept. Progr Energy Combust Sci 38:522–550

Micklander E, Bertram HC, Marno H et al (2005) Early post-mortem discrimination of water-holding capacity in pig longissimuss muscle using new ultrasound method. Food Sci Technol 38:437–445

Milewski S, Zaleska B (2011) The effect of dietary supplementation with *Saccharomyces cerevisiae* dried yeast on lambs' meat quality. J Anim Feed Sci 20:537–545

Miller TL, Wolin MJ (2001) Inhibition of growth of methane-producing bacteria of the ruminant forestomach by hydroxymethylglutaryl-SCoA reductase inhibitors. J Dairy Sci 84:14451448

Moukadiri I, Armero J, Abad A et al (1997) Identification of a mannoprotein present in the inner layer of the cell wall of *Saccharomyces cerevisiae*. J Bacteriol 179:2154–2162

Nagarawa TG, Newbold CJ, Van Nevel CJ (1997) Manipulation of ruminal fermentation. In: Hobson PN, Stewart CS (eds) The rumen microbial ecosystem, 2nd edn. Chapman & Hall, London, pp 523–632

Nakagawa T, Kaichiro Y, Tatsuro M et al (2002) Cold-active pectinolytic activity of psychrophilic-basidiomycetous yeast *Cystofilobasidium capitatum* strain PPY-1. J Biosci Bioeng 94:175–177

Nocek JE, Holt MG, Oppy J (2011) Effects of supplementation with yeast culture and enzymatically hydrolyzed yeast on performance of early lactation dairy cattle. J Dairy Sci 94:4046–4056

Nudda A, Battacone G, Bencini R, Pulina G (2004) Nutrition, and milk quality. In: Pulina G (ed) Dairy sheep nutrition. CABI Publishing, Wallingford, pp 129–149

Oliveira ACD, Fernandes ML, Mariano AB (2014) Production and characterization of an extracellular lipase from *Candida guilliermondii*. Braz J Microbiol 45:1503–1511

Ouellet DR, Chiquette J (2016) Effect of dietary metabolizable protein level and live yeasts on ruminal fermentation and nitrogen utilization in lactating dairy cows on a high red clover silage diet. Anim Feed Sci Technol 220:73–82

Owens FN, Secrist DS, Hill WJ et al (1998) Acidosis in cattle: a review. J Anim Sci 76:275–286

Panda AK, Singh R, Pathak NN (1995) Effect of dietary inclusion of *Saccharomyces cerevisiae* on growth performance of crossbred calves. J Appl Anim Res 7:195–200

Paryad A, Mahmoudi M (2008) Effect of different levels of supplemental yeast (*Saccharomyces cerevisiae*) on performance: blood constituents' and carcass characteristics of broiler chicks. Afr J Agric Res 3:835–842

Patra K (2013) The effect of dietary fats on methane emissions, and its other effects on digestibility, rumen fermentation and lactation performance in cattle: a meta-analysis. Livestock Sci 155:244–254

Pereira-Meirelles FV, Rocha-Leao MHM, Sant'Anna GL Jr (2000) Lipase location in *Yarrowia lipolytica* cells. Biotechnol Lett 22:71–75

Pinloche E, McEwan N, Marden J-P et al (2013) The effects of a probiotic yeast on the bacterial diversity and population structure in the rumen of cattle. PLoS One 8:e67824. https://doi.org/10.1371/journal.pone.0067824

Pinos-Rodriguez JM, Robinson PH, Ortega ME et al (2008) Performance and rumen fermentation of dairy calves supplemented with *Saccharomyces cerevisiae*1077 or *Saccharomyces boulardii*1079. Anim Feed Sci Technol 140:223–232

Poondla VK, Yannam SK, Gummadi SN et al (2016) Enhanced production of pectinase by *Saccharomyces cerevisiae* isolate using fruit and agroindustrial wastes: its application in fruit and fiber processing. Biocatal Agric Biotech 6:40–50

Poppy GD, Rabiee AR, Lean IJ et al (2012) A meta-analysis of the effects of feeding yeast culture produced by anaerobic fermentation of *Saccharomyces cerevisiae* on milk production of lactating dairy cows. J Dairy Sci 95:6027–6041

Prandini A, Tansini G, Sigolo S et al (2009) On the occurrence of aflatoxin M1 in milk and dairy products. Food Chem Toxicol 47:984–991

Preston TR, Leng R (1987) Digestive physiology of ruminants. In: Matching ruminant with available resources in the tropics and sub-tropics. https://www.cipav.org.co/PandL/Preston_Leng.htm

Ramos GRV, Birchal VS, Seara LM et al (2011) Chemical characterization of yeast autolysate and assessment of the acceptability of cheese buns containing dried autolysate. Rev Nutr 24:473–484

Ratledge C (2004) Fatty acid biosynthesis in microorganisms being used for single cell oil production. Biochimie 86:807–815

Ratledge C, Wynn JP (2002) The biochemistry and molecular biology of lipid accumulation in oleaginous microorganisms. Adv Appl Microbiol 51:1–52

Rattray JBM (1988) Yeasts. In: Microbial lipids. Academic, London, pp 555–697

Rossow HA, Riordan T, Riordan A (2018) Effects of addition of a live yeast product on dairy cattle performance. J Appl Anim Res 46:159–163

Rufino LDA, Pereira OG, Ribeiro KG et al (2013) Effect of substitution of soybean meal for inactive dry yeast on diet digestibility, lamb's growth, and meat quality. Small Ruminant Res 111:56–62

Russell JB (2002) Rumen microbiology and its role in ruminant nutrition. Russell Publ, Ithaca

Sakata T, Tamate H (1978) Rumen epithelial cell proliferation accelerated by rapid increase in intraruminal butyrate. J Dairy Sci 61:1109–1113

Salihu A, Alama MZ, AbdulKarim MI et al (2012) Lipase production: an insight in the utilization of renewable agricultural residues. Resour Conserv Recycl 58:36–44

Santra A, Karim SA (2003) Rumen manipulation to improve animal productivity. J Anim Sci 16:748–763

Sedlak M, Ho NWY (2004) Production of ethanol from cellulosic biomass hydrolysates using genetically engineered *Saccharomyces* yeast capable of cofermenting glucose and xylose. Appl Biochem Biotechnol 113:403–416

Sitepu IR, Sestric R, Ignatia L et al (2013) Manipulation of culture conditions alters lipid content and fatty acid profiles of a wide variety of known and new oleaginous yeast species. Bioresour Technol 144:360–369

Souza NE, Visentainer JV (2006) Colesterol da mesa ao corpo. Livraria Varela, São Paulo

Souza XR, Bressan MC, Perez JRO et al (2004) Efeitos do grupo genético, sexo y peso ao abate sobre as propriedades fisicoquímicas da carne de cordeiros em crescimento. Ciênc Tecnol Alimentos 24:543–549

Steinfeld H, Gerber P, Wassenaar T et al (2006) Livestock's role in climate change and air pollution. In: Livestock long shadow: environmental issues and options, FAO report, Rome, pp 79–124

Titi HH, Dmour RO, Abdullah AY (2008a) Growth performance and carcass characteristics of Awassi lambs and Shami goat kids fed yeast culture in their finishing diet. Anim Feed Sci Tech 142:33–43

Titi HH, Abdullah AY, Lubbadeh WF et al (2008b) Growth and carcass characteristics of male dairy calves on a yeast culture-supplemented diet. South Afr J Anim Sci 38:174–183

Tripathi MK, Karim SA (2010) Effect of individual and mixed live yeast culture feeding on growth performance, nutrient utilization, and microbial crude protein synthesis in lambs. Anim Feed Sci and Technol 155:163–171

Tripathi MK, Karim SA (2011) Effect of yeast cultures supplementation on live weight change, rumen fermentation, ciliate protozoa population, microbial hydrolytic enzymes status and slaughtering performance of growing lamb. Livestock Sci 135:17–25

Vakhlu J, Kour A (2006) Yeast lipases: enzyme purification, biochemical properties, and gene cloning. Eur J Biotechnol 9:69–81

Valadares Filho SC, Machado PAS, Chizzotti ML, et al (2010) CQBAL 3.0 Tabelas brasileiras de composicao de alimentos para bovinos. http://cqbal.agropecuaria.ws/index.php

Visintin S, Ramos L, Batista N et al (2017) Impact of *Saccharomyces cerevisiae* and *Torulaspora delbrueckii* starter cultures on cocoa beans fermentation. Int J Food Microbiol 257:31–40

Vohra A, Satyanarayana T (2001) Phytase production by the yeast *Pichia anomala*. Biotechnol Lett 23:551–554

Vohra A, Syal P, Madan A (2016) Probiotic yeasts in livestock sector. Anim Feed Sci Technol 2191:31–47

Vyas S, Chhabra M (2017) Isolation, identification, and characterization of *Cystobasidium oligophagum* JRC1: a cellulase and lipase producing oleaginous yeast. Bioresour Technol 223:250–258

Wallace RJ, McKain N, Broderick GA et al (1997) Peptidases of the rumen bacterium *Prevotella ruminicola*. Anaerobe 3:35–42

Wang LZ, Zhou ML, Wang JW et al (2016) The effect of dietary replacement of ordinary rice with red yeast rice on nutrient utilization, enteric methane emission and rumen archaeal diversity in goats. PLoS One 11:160–198

Wanlapa L, Ancharida A, Motofumi S et al (2013) Diversity and fermentation products of xylose-utilizing yeasts isolated from buffalo feces in Thailand. Microb Environ 28:354–360

Westhoek H, Rood T, van den Berg M et al (2011) The protein puzzle. PBL Netherlands Environmental Assessment Agency, The Hague, p 218

Whitehead DC (1995) Grassland nitrogen. CAB International, Wallingford

Williams PEV, Tait CAG, Innes GM et al (1991) Effect of inclusion of yeast culture (*Saccharomyces cerevisiae* plus growth medium) in the diet on milk yield and forage degradation pattern in the rumen of sheep and steer. J Anim Sci 69:3016–3026

Xiao JX, Alugongo GM, Chung R et al (2016) Effects of *Saccharomyces cerevisiae* fermentation products on dairy calves: ruminal fermentation, gastrointestinal morphology, and microbial community. J Dairy Sci 99:5401–5412

Yalçın S, Yalçın S, Can P et al (2011) The nutritive value of live yeast culture (*Saccharomyces cerevisiae*) and its effect on milk yield, milk composition and some blood parameters of dairy cows. Asian-Aust J Anim Sci 24:1377–1385

Zain M, Jamarun N, Arnim A et al (2011) Effect of yeast (*Saccharomyces cerevisiae*) on fermentability, microbial population and digestibility of low-quality roughage *in vitro*. Archiv Zootechnica 14:51–58

Yeast Culture and Direct-Fed Microbes: Modes of Action and Beneficial Applications in Ruminants

Wen Zhu and Jian-xin Liu

Abstract Improvements in animal feed utilization, production, health, and food safety are the most important goals of ruminant nutrition studies. Several feed additives have been used to achieve these goals in practice. After banning the use of antibiotics in animal feed and in the absence of a suitable growth stimulant, positive manipulation of yeast products has become a popular field of research. Yeast culture and direct-fed microbes are two main categories of yeast products that have become commercially available. These yeast products have been documented to improve ruminant growth, productivity, immunity, and health and modify rumen fermentation depending on the different modes of action. This chapter summarizes the definition, yeast product mechanism of action, and yeast culture supplementation effect on animal performance, rumen fermentation and microbiota, and immune response.

1 Introduction

Yeast products are widely used as rumen fermentation modifiers in modern dairy farms. Currently, two main categories of yeast products are commercially available (AAFCO 2011). The American AAFCO (2011) indicates that the first group consists of yeast culture (YC) produced through yeast fermentation. This type usually consists of three components: (1) yeast cells; (2) media consisting of ground corn, hominy feed, corn gluten feed, wheat middling, rye middlings, diastic malt, corn syrup, and cane molasses; and (3) fermentation by-products. Yeast cultures are not dependent on live yeast for their physiological effects. Many studies have been conducted on YC, such as *Saccharomyces cerevisiae*, *S. boulardii*, *Trichosporom*

W. Zhu
College of Animal Science and Technology, Anhui Agricultural University, Hefei, People's Republic of China

J.-x. Liu (✉)
College of Animal Sciences, Zhejiang University, Hangzhou, People's Republic of China
e-mail: liujx@zju.edu.cn

A. Z. M. Salem et al. (eds.), *Exogenous Enzymes as Feed Additives in Ruminants*, https://doi.org/10.1007/978-3-031-27993-5_2

sericeum, and *Aspergillus oryzae*. Among them, *S. cerevisiae* has the most potential as a growth-promoting feed additive in ruminants (Tripathi and Karim 2010).

In contrast, the second group consists of active dry yeast products (AAFCO 2011) that contain more than 15 billion live yeast cells per gram by definition. Varieties of direct-fed microbials (DFMs) are being used in ruminants around the world and can be classified into three major classes: (1) bacterial, (2) fungal, and (3) bacterial and fungal combinations (Khan et al. 2016). Generally, there are three types of DFMs used in ruminants: rumen microbes, lactic-acid-producing bacteria (LAB), and other bacteria. Logically, DFMs derived from the rumen may be more capable of adjusting to the microbial community, and the effects depend on the yeast cell being alive in the rumen to have a production effect (Dawson et al. 1990; Newbold et al. 1996). The population of rumen microbes is composed of different kinds of bacteria, protozoa, and fungi. However, only a small number of these communities qualify based on their potential as DFMs (Miron et al. 2001). The majority of DFMs used in ruminant production are LAB, followed by lactic-acid-utilizing bacteria (LUB) and other microorganisms (Elghandour et al. 2015). Some other types of bacteria, such as *Bifidobacterium* spp. and *Bacillus* spp., have also been successfully used as DFMs. The most common genera used as DFMs include *Propioni bacterium*, *Streptococcus*, *Lactobacillus*, *Bacillus* and some other important bacteria, such as *Prevotella bryantii* and *Megasphaera elsdenii* (*M. elsdenii*) (Kung 2006; Seo et al. 2010). *S. cerevisiae* has been extensively used as a DFM for improved performance and milk yield in dairy cattle. In addition to *S. cerevisiae*, *Aspergillus oryzae* and *A. niger* have also been used in ruminant nutrition (McAllister et al. 2011). The aim of this chapter is to comprehensively describe the mechanism of action of YC and DFM as well as their recent advances in ruminants.

2 Mechanism of Action of Yeast Culture and Direct-Fed Microbes

2.1 Mechanism of Action of Yeast Culture

Due to the differences between YC and DFM, the active ingredients and putative modes of action of these two products differ. Yeast culture contains yeast cells and compounds produced during fermentation activity, such as B vitamins, organic acids, and polyphenols, and all of them might be responsible for the positive effects on performance and health when incorporated into the diet of ruminants. Components of the yeast cell wall, such as oligosaccharides glucan and mannan, might benefit local and systemic immune responses (Newman 1994; Murphy et al. 2007). Furthermore, metabolites produced by *S. cerevisiae* fermentation might have potential antimicrobial activities against pathogens (Jensen et al. 2008), and these soluble products not only inhibit the growth of *Escherichia coli* but also suppress the activity of the bacteria. Metabolites present in YC have been shown to modulate immune response in vitro, with a reduction in inflammatory response and oxidative stress (Jensen et al. 2007).

Yeast culture products contain compounds that favorably alter ruminal microbial fermentation by stimulating the growth and activity of fiber-digesting bacteria (Harrison et al. 1988; Mao et al. 2013), increasing fiber digestibility (Yoon and Garrett 1998), increasing microbial crude protein (MCP) synthesis (Hristov et al. 2010), stimulating the growth of lactate-utilizing bacteria, and decreasing the accumulation of lactate (Callaway and Martin 1997). Branched-chain fatty acids or B vitamins from yeast culture may also provide stimulatory factors for cellulolytic bacteria (Weidmeier et al. 1987).

2.2 Mechanism of Action of Direct-Fed Microbes

Many theories have been proposed to explain the mechanism through which DFM exerts its beneficial impacts. Moreover, some DFMs act via the rumen, while others influence the gastrointestinal tract (Puniya et al. 2015). Inside the rumen, LAB, a type of DFM, are environmentally friendly, and these bacteria produce lactic acid, which is one of the key compounds that can change the pH of the competitor bacteria. LAB, such as *Lactobacilli* and *Enterococci*, might prevent ruminal acidosis in dairy cows (Nocek et al. 2011) by facilitating the growth of ruminal microorganisms to adapt to the presence of lactic acid in the rumen (Yoon and Stern 1996) and by stimulating LUB. In addition, DFM produces an antimicrobial substance known as bacteriocin (Servin 2004; McAllister et al. 2011). The rumen pH is mainly determined by the lactic acid concentration (Williams et al. 1991). The LUB has been successfully used as a DFM to decrease lactate concentrations and keep ruminal pH within a normal range. For example, *M. elsdenii* utilizes lactate and prevents drastic decreases in pH caused by the accumulation of lactate in the rumen when fed with a highly fermentable diet (Kung and Hession 1995). Propionibacteria ferment lactate to propionate. Since propionate is the major precursor for gluconeogenesis during early lactation in dairy cows (Reynolds et al. 2003), enhanced propionate in the rumen can increase glucose production (Stein et al. 2006). Usually, propionibacteria are present in the animal diet at medium concentrations, which is expected to convert lactate into propionate in the rumen, resulting in an increased glucogenic potential for milk production (Stein et al. 2006). Fungal DFM has been extensively used in ruminants for improving forage digestion and animal performance. Rose et al. (1986) suggested that yeasts remove oxygen in the rumen. Yeast cells in the rumen use available oxygen on the surfaces of freshly ingested feed to maintain metabolic activity. In addition, DFM can compete with starch-utilizing bacteria, thereby preventing the accumulation of lactate and providing vitamins and organic acids to cellulolytic bacteria (Chaucheyras et al. 1995; Lynch and Martin 2002).

Dietary supplementation with DFM showed beneficial effects on the intestinal tracts. The potential mechanisms are summarized as follows: (1) production of antibacterial compounds (acids, bacteriocins, and antibiotics); (2) competition with pathogens for mucosa colonization and/or nutrients; (3) production and/or stimulation of enzymes; (4) stimulation of immune response by the host; and (5) metabolism and detoxification of desirable compounds (Seo et al. 2010). In addition, DFM can modulate the host immune system by stimulating and activating dendritic cells, natural killer cells, macrophages, T and B lymphocytes, and neutrophils of the intestines (Krehbiel et al. 2003).

3 Beneficial Applications of Yeast Culture and Direct-Fed Microbes

The beneficial effects of adding YC and DFM to the diets of ruminants include the effects on animal performance, rumen fermentation, microbial communities, and host immunomodulation.

3.1 Effects of Yeast Culture and Direct-Fed Microbes on Animal Performance

3.1.1 Effect on Growth

Neonatal calves often face various stresses, such as transportation, weaning, vaccination, and dehorning (Krehbiel et al. 2003). In modern intensive farming, calves are separated from cows before their intestinal microbiota have completely colonized. This situation might increase the possibility of diarrhea and weight loss (Seo et al. 2010). The administration of beneficial microorganisms in calves may speed up the colonization of microbiota in the intestine and enhance the return of gut intestinal function back to normal (Kung Jr 2001). Therefore, many studies have been conducted to evaluate the effects of yeast products on young calves.

Variable responses of feeding yeast products on the performance of young calves have been reported. According to Lesmeister et al. (2004), the addition of 1% YC to the grain of young calves did not affect performance, while the addition of 2% YC improved the dry matter intake (DMI), body weight (BW) gain, and feed efficiency compared with a diet without YC. However, Magalhães et al. (2008) did not find any positive effects of YC addition on grain intake and body weight gain. The reasons for the variations in the responses to yeast supplementation were related to diet composition, level of yeast feeding, yeast strain and supplementation method (Pinos-Rodríguez et al. 2008). An increase in rumen propionate has been observed in many previous studies on diets supplemented with YC (Harrison et al. 1988; Plata et al. 1994; Dias et al. 2018), and higher propionate concentrations would lead to increased glycogenicity in ruminants. Thus, YC is expected to generally increase the efficiency of energy utilization for fattening. This phenomenon may be desirable in beef production.

Many studies indicate that LAB could improve weight gain and feed efficiency when used as a DFM source (Seo et al. 2010). Abe et al. (1995) investigated the effects of oral administration of *Bifidobacterium pseudolongum* and *Lactobacillus acidophilus* on newborn calves. The BW gain was different between the treated and control groups, although no difference was found between the groups fed with *Bifidobacterium* and *Lactobacillus*. Galvão et al. (2005) reported improvements in grain intake, BW gain, and blood parameters of calves fed with live yeast incorporated into the grain during the preweaning period but not during the postweaning period. Moreover, LAB were successfully inoculated into young calves to improve growth performance (Frizzo et al. 2010).

Most bacterial DFMs for young calves contain LAB, whereas dairy propionibacteria are rarely used. Propionibacteria can stimulate rumen development by increasing the concentrations of propionate and butyrate (Seo et al. 2010). Adams et al. (2008) estimated the effect on calf growth performance of a novel bacterial strain, *Propionibacterium jensenii* 702, isolated in Australia. Fecal recovery of *P. jensenii* 702 from the treatment groups at week 2 indicated successful gastrointestinal transit of the bacterium, and the treated calves exhibited higher BW gain during the pre- and postweaning periods.

3.1.2 Effect on Intake and Lactation Yield

Modern dairy farms usually utilize diets with high nutrient concentrations to meet the metabolic demand of high milk production. Such a feeding system is associated with metabolic dysfunction, such as subacute rumen acidosis, especially during poor feeding conditions. The transition period, defined as 3 weeks prepartum to 3 weeks postpartum, is a critical period in dairy cows because the increased demand for nutrients and energy exceed dry matter intake, resulting in tremendous physiological challenges to maintain homeostasis at the onset of lactation (Overton and Waldron 2004). The physiological challenges include immune function, which is suppressed in the transition period, resulting in increased disease susceptibility (Nonnecke et al. 2003; Ohtsuka et al. 2006). Thus, before parturition and extending through peak lactation, the lactating cows are supplemented with yeast products.

Effects of YC

Feeding 56–60 g/d of *S. cerevisiae* fermentation product during the transition period increased DMI and milk yield by 2–5% on average (Robinson and Garrett 1999; Dann et al. 2000). In contrast, Bruno et al. (2009) and Hristov et al. (2010) reported no significant difference in DMI in lactating Holstein cows between the YC-supplemented and control groups. In a study by Zhu et al. (2016), supplementation with YC had no influence on the DMI in lactating dairy cows under heat stress. Schingoethe et al. (2004) observed a decrease in DMI with YC supplementation. In summary, the effect of YC supplementation on DMI is variable. The possible sources of variation in response to YC supplementation may be the type and level of YC used (Di Francia et al. 2008; Poppy et al. 2012). According to Poppy et al. (2012), the variation was due to the interaction between the YC and lactation stage. Cows in early lactation (<70 days in milk (DiM)) had increased DMI with YC supplementation, while cows in later lactation (>70 DiM) had decreased DMI. Supplementing YC decreased DMI linearly with increasing doses of YC in mid-lactation cows, with an average DiM of 180 (Zhu et al. 2017). Zaworski et al. (2014) observed higher milk yield with no change in milk composition with YC supplementation during the transition period in dairy cows. Several investigators focused on dairy animals (Dann et al. 2000; Nocek et al. 2003) have reported increased prepartum and postpartum DMI and milk production after dietary supplementation with YC. Bruno et al. (2009) reported that lactating cows receiving YC during heat stress produced 1.2 kg/d more milk and more milk protein, solids-not-fat, and lactose than the control cows, although energy-corrected milk (ECM) yield and milk protein percentage did not differ between the treatments.

Yeast culture may have beneficial functions in stressed animals. Feeding YC improved the milk yield and milk components in heat-stressed multiparous Holstein cows (Bruno et al. 2009). Zhu et al. (2016) reported increased milk yields with YC supplementation in dairy cows under heat stress. However, reports have also shown that the milk yield and milk composition were not affected by supplementation with YC (Hristov et al. 2010). Feeding YC improved the yields of milk, 3.5% fat-corrected milk (FCM), and milk components, and the increments were not affected based on whether the cows were fed with low or high starch (Dias et al. 2018). Feeding YC may improve productive performance of dairy cows in numerous experiments, and a meta-analysis of the effects of YC showed increased yields of milk and milk components irrespective of the method of adding YC (either mixed with total mixed ration (TMR) or fed as top-dressing; Poppy et al. 2012). According to the meta-analysis results (Poppy et al. 2012), the percentage of milk fat and milk protein was not affected by YC supplementation.

Effects of DFM

Numerous experiments have reported that DFMs have positive effects on the DMI, milk yield, and milk composition in lactating dairy cows during the transition period (Nocek and Kautz 2006; Ferguson et al. 2010; Qiao et al. 2010). Most of these studies were performed using LAB, LUB, yeast, or fungi as DFMs (Nocek et al. 2003; Oetzel et al. 2007; Weiss et al. 2008). Treatment of alfalfa with *L. acidophilus* improved milk yield and composition in dairy cows with no effect on DMI (Colenbrander et al. 1988). Higher milk production (0.73 kg/day) with less DMI (0.42 kg/day) was observed when cows were treated with lactic acid bacteria, including *L. casein*, *L. acidophilus* and *Enterococcus faecium* in comparison with the control (Gomez-Basauri et al. 2001). Using *L. acidophilus* NP51 and *Propionibacterium freudenreichii* NP24 improved milk yield in heat-stressed cows (Boyd et al. 2011). Other studies have shown higher milk yield when lactating cows were supplemented with a combination of lactic acid bacteria acidophilus (*L. plantarum*) and/or *E. faecium* and fungal culture (*S. cerevisiae*) (Block et al. 2000). Milk production in lactating cows increased when *M. elsdenii* NCIMB 41125 was supplemented compared with the control (Aikman et al. 2008). Differences were not observed in the 3.5% FCM, milk fat yield or milk protein percentage and yield in pre- and postpartum dairy cows treated with Biomate yeast combined with two strains of *E. faecium* compared with that of the controls (Nocek and Kautz 2006). In agreement, Luan et al. (2015) also reported that neither DMI nor milk yield was affected in dairy cows supplemented with *Bacillus pumilus* 8G-134 from 21 days before the expected calving date to 154 days after calving.

3.2 Effects on Rumen Fermentation

Ruminal health is the basis for the health of dairy cows. Yeast products are thought to improve animal performance by influencing the rumen microbiota in a way that support the growth and activity of microorganisms.

3.2.1 Effects of YC

Yeast culture is known to stabilize ruminal pH by reducing the lactic acid concentration of the ruminal fluid (Williams et al. 1991). The YC of *T. sericeum* was reported to significantly increase the rumen pH in sheep fed with a hay and concentrate diet due to the stabilizing effect of YC on pH by reducing the concentration of lactate (Mwenya et al. 2004). Ammonia is the main source of nitrogen (N) for MCP synthesis (Bach et al. 2005). Moloney and Drennan (1994) reported no effect of YC on rumen ammonia-N in steers fed with a low protein diet, while YC significantly decreased rumen ammonia in the study of Harrison et al. (1988). An increase in the overall utilization of ammonia-N and increased MCP synthesis with the addition of YC have been reported in many studies (Hristov et al. 2010; Mao et al. 2013). The increase in ruminal total volatile fatty acid (VFA) concentration in response to YC supplementation was indicative of stimulating rumen microbial activities. In vivo and in vitro studies have documented the positive effects of YC on rumen fermentation (Yoon and Stern 1996; Mao et al. 2013). Increases in rumen propionate have also been observed in previous studies focused on YC supplementation (Harrison et al. 1988; Yoon et al. 1997). Similar results were observed when YC was fed to cows on low-quality forage diets (Zhu et al. 2017).

3.2.2 Effects of DFM

The strategic use of DFMs with feed reduces the risk of subclinical ruminal acidosis in feedlots. A study by Kung and Hesion (1995) reported that *M. elsdenii* significantly prevented a drop in pH in vitro with diets containing readily fermentable carbohydrate. Similar results were found in Angus crossbred steers treated with *M. elsdenii* (McDaniel 2009). Nocek et al. (2003) observed a reduced area under pH 5.5 for dairy cows receiving *Enterococcus* and *Lactobacillus* compared to cows fed with a control diet. In contrast, treatment with *M. elsdenii* had no effect on rumen pH in lactating cows on a high concentrate diet (Hagg 2007). No significant effect of LAB (*Lactobacillus mesenteroides*) on rumen ammonia-N was observed in sheep fed with a hay and concentrate diet (Mwenya et al. 2004). The ammonia-N concentration in the rumen of lactating cows treated with *Prevotella bryantii* was greater than that in the controls (Chiquette et al. 2008), which was probably due to greater proteolytic and deamination activity by *P. bryantii* (Chiquette et al. 2008). Concentrations of rumen VFA and lactate in lactating cows on high concentrate diets were not affected by treatment with *M. elsdenii* (Hagg 2007). However, positive effects of feeding *S. cerevisiae* on ruminal fermentation have been reported in many other studies (Pinos-Rodríguez et al. 2008; Qiao et al. 2010). A study by Beauchemin et al. (2003) observed that *E. faecium* significantly increased propionate and consequently tended to decrease butyrate concentrations in feedlot cattle due to stimulation of LUB to produce propionate. The bacterium *Chryseobacterium kwangyangense* significantly increased the production of acetate, propionate, and butyrate from rice straw and oil palm fronds compared with *Enterobacter cloacae* and *E. aerogenes* in an in vitro study by Ramin et al. (2008). Acetate, butyrate, and branched-chain fatty acids were greater postfeeding in early-lactation dairy cows with *P. bryantii* supplementation (Chiquette et al. 2008). The increased concentrations of acetate and butyrate suggest an increased rate of fiber fermentation, and the enhanced branched-chain fatty acids indicate an increased rate of proteolysis and amino acid metabolism in treated cows (Chiquette et al. 2008). Kenney et al. (2015) reported that the LAB DFM did not modulate ruminal fermentation but shifted the ruminal VFA profile and pH in steers fed with a high-concentrate diet.

3.3 *Effects on Rumen Microbiota*

The response in terms of VFA production and patterns with yeast product supplementation is a consequence of the effects of these products on the growth of different species of rumen microbes.

3.3.1 Effects of YC

Newbold et al. (1998) reported a 25% increase in total bacterial counts with *S. cerevisiae* YC in vitro. Supplementation with YC increased the rumen fungal population (Mao et al. 2013; Zhu et al. 2017), and stimulation of the growth of cellulolytic bacterial populations (*Ruminococcus albus*, *R. flavefaciens* and *Fibrobacter succinogenes*) has been reported in many studies (Callaway and Martin 1997; Zhu et al. 2017). A number of experiments have focused on the effect of YC on the protozoa population. In some cases (Yoon and Stern 1996; Mwenya et al. 2005), YC did not modify protozoa numbers, although in other cases, the protozoa number was increased (Mao et al. 2013; Zhu et al. 2017). Arakaki et al. (2000) reported that steers supplemented with *S. cerevisiae* fermentation product (SCFP) decreased the proportion of *Entodinium* from 87.7% to 69.6%. *Entodinium* is a genus of rumen protozoa responsible for engulfing bacteria and reducing the microbial protein supply to the small intestine (Ivan et al. 2000).

3.3.2 Effects of DFMs

Ghorbani et al. (2002) reported a significant increase in protozoa number with a concomitant increase in rumen ammonia concentration due to protozoa recycling bacterial-N in feedlot cattle treated with *Propionibacterium*, with most of the protozoal populations identified as *Entodinium.* Hristov et al. (2010) demonstrated that a large population of *Entodinium* persisted in the rumen of cattle fed with high-grain diets based on barley. The number of amylolytic bacteria decreased with *Propionibacterium* treatment because of predation from protozoa (Ghorbani et al. 2002). The total number of bacterial genomes was unaffected by *M. elsdenii* in steers fed with high concentrate diets, while populations of *M. elsdenii* were significantly increased (McDaniel 2009). Inoculation with *M. elsdenii* can be functional in increasing rumen populations of LUB (McDaniel 2009). Ruminal LAB numbers were significantly greater in control and *E. faecium* treatments than in *Propionibacterium* and *E. faecium* combined with yeast culture (Yang et al. 2004).

3.4 Beneficial Effects of Yeast Culture and Direct-Fed Microbes on Host Immunomodulation

The benefits of yeast product feeding were based on the potential for a positive intestinal effect, including the establishment of a desirable gastrointestinal microflora and the prevention of pathogenic bacterial colonization and stimulation of the immune system.

3.4.1 Effects of YC

Pathogenic bacteria that attach to the host can cause diarrhea in calves. Previous research has demonstrated that feeding YC might have the potential to decrease the risk of diarrhea. For example, mannan oligosaccharides can act as antiadhesive agents and prevent the colonization of pathogens in the intestinal tract (Ganner and Schatzmayr 2012). Furthermore, the metabolites present in YC can minimize the growth of enteric pathogens (Jensen et al. 2008) or reduce the inflammatory response in the gut (Jensen et al. 2007). The majority of *Salmonella* spp. isolates adhered to the cell wall of *S. cerevisiae* (Pérez-Sotelo et al. 2005), which might prevent the attachment and invasion of intestinal cells. Heinrichs et al. (2003) reported that adding mannan-oligosaccharides to milk replacer fed to calves improved fecal scores. Although a similar measure of systemic immunity was observed between calves fed with YC and those that served as controls, costs associated with treatments to restore health were lower for calves fed with YC (Magalhães et al. 2008), suggesting that feeding YC might improve the overall gut health of calves. Franklin et al. (2005) reported that supplementing mannan-oligosaccharides to cows during the last 3 weeks of the dry period enhanced their antibody response to rotavirus and tended to enhance the subsequent transfer of rotavirus antibodies to calves. Nocek et al. (2011) reported that supplementation with YC throughout the first 14 weeks of lactation decreased somatic cell counts and cases of clinical mastitis. Similar results were observed by Zaworski et al. (2014), who found that YC supplementation during the transition period decreased somatic cell count in early lactation. Yuan et al. (2015) reported that supplementation with YC enhanced measures of humoral and mucosal immunity and modulated uterine inflammatory signals and mammary gland health in transition dairy cows.

3.4.2 Effects of DFMs

Bacillus species have the potential to produce antimicrobial compounds and have shown potential benefits in immune modulation in ruminants (Sun et al. 2010; Novak et al. 2012). DFM stimulates epithelial innate immunity, which may suppress intestinal inflammation by increasing the production of TNF-α (Pagnini et al. 2010).

The production of cytokines and chemokines in addition to T- and B-cell responses influenced by DFMs depends on the types of DFM and the dose and duration of the experiment (McAllister et al. 2011). Abu-Tarboush et al. (1996) observed that calves fed with *L. acidophilus* 27SC had a significantly lower incidence of diarrhea compared with the controls, which confirmed the positive effect of *Lactobacilli* in decreasing the incidence of calf diarrhea. Similar results were observed by Kim et al. (2011), who found that feeding multispecies DFM (combination of three LAB and three *Bacillus* strains) during the preweaning period could reduce calf diarrhea compared with the calves fed with antibiotics. At birth, neonatal calves do not have immunoglobulins (IgG and IgA) in their blood streams and rely on colostrum for immunity; thus, passive immunity is important in the disease control of neonatal calves before weaning (Wittum and Perino 1995). Sun et al. (2010) reported a change in immunoglobulins associated with a *Bacillus* DFM fed to neonatal calves, and they observed elevated levels of serum IgG in DFM-treated calves but no change in IgA or IgM levels. In contrast, Luan et al. (2015) observed increased levels of milk IgA in DFM-treated calves and did not find a difference in IgG or IgM. In a study by Novak et al. (2012), administration of a *Bacillus* DFM to dairy calves resulted in altered immune cell populations and an improvement in immune development.

4 Conclusion and Prospects

Manipulation of gastrointestinal microbial ecosystems to augment animal performance and safeguard health is one of the important goals of animal scientists. Yeast products are among the feed additives used worldwide. This chapter identified a positive ruminant performance response with the supplementation of yeast culture on various aspects of health. In addition, the chapter also discussed the inconsistency in the response of animals, which may depend on the dose and diet dependence of yeast culture effects, along with animal factors such as age, physiological stage, health, and stress status. In view of the increasing concern about the stress dairy cattle suffer, yeast culture research in the future should be directed toward dairy cows in the transition period or under heat stress.

References

AAFCO (2011) Official Publication. Association of the American Feed Control Officials, Oxford

Abe H, Funada R, Ohtani J et al (1995) Changes in the arrangement of microtubules and microfibrils indifferentiating conifer tracheids during expansion of cells. Ann Bot 75:305–310

Abu-Tarboush HM, Al-Saiady MY, El-Din AHK (1996) Evaluation of diet containing lactobacilli on performance, fecal coliform and lactobacilli of young dairy calves. Anim Feed Sci Technol 57:39–49

Adams MC, Luo J, Rayward D et al (2008) Selection of a novel direct-fed microbial to enhance weight gain in intensively reared calves. Anim Feed Sci Technol 145:41–52

Aikman PC, Henning PH, Jones AK et al (2008) Effect of administration of *Megasphaeraelsdenii* NCIMB 41125 lactate utilizing bacteria in early lactation on the production, health and rumen environment of highly productive dairy cows fed a high concentrate diet. KK Animal Nutrition Internal Report, KKAN, Umbogintwini

Arakaki LC, Stahringer RC, Garrett JE et al (2000) The effects of feeding monensin and yeast culture, alone or in combination, on the concentration and generic composition of rumen protozoa in steers fed on low-quality pasture supplemented with increasing levels of concentrate. Anim Feed Sci Technol 84:121–127

Bach A, Calsamiglia S, Stern MD (2005) Nitrogen metabolism in the rumen. J Dairy Sci 88(E Suppl):E9–E21

Beauchemin KA, Yang WZ, Morgavi DP et al (2003) Effects of bacterial direct-fed microbials and yeast on site and extent of digestion, blood chemistry, and subclinical ruminal acidosis in feedlot cattle. J Anim Sci 81:1628–1640

Block E, Nocek JE, Kautz WP et al (2000) Direct fed microbial and anionic salt supplementation to dairy cows fed 21 days pre- to 70 days postpartum. J Anim Sci 78:304–304

Boyd J, West JW, Bernard JK (2011) Effects of the addition of direct-fed microbials and glycerol to the diet of lactating dairy cows on milk yield and apparent efficiency of yield. J Dairy Sci 94:4616–4622

Bruno RGS, Rutigliano HM, Cerri RL et al (2009) Effect of feeding *Saccharomyces cerevisiae* on performance of dairy cows during summer heat stress. Anim Feed Sci Technol 150:175–186

Callaway ES, Martin SA (1997) Effects of a *Saccharomyces cerevisiae* culture on ruminal bacteria that utilize lactate and digest cellulose. J Dairy Sci 80:2035–2044

Chaucheyras F, Fonty G, Bertin G et al (1995) Effects of live Saccharomyces cerevisiae cells on zoospore germination, growth and cellulolytic activity of the rumen anaerobic fungus, *Neocallimastix frontalis* MCH3. Curr Microbiol 31:201–205

Chiquette J, Allison MJ, Rasmussen MA (2008) *Prevotellabryantii*25A used as a probiotic in early-lactation dairy cows: effect on ruminal fermentation characteristics, milk production, and milk composition. J Dairy Sci 91:3536–3543

Colenbrander VF, Grant RJ, Schaaf G (1988) Milk production and feed intake of dairy cows fed *lactobacillus* inoculated alfalfa silage. Appl Agric Res 3:55–59

Dann HM, Drackley JK, McCoy GC et al (2000) Effects of yeast culture (*Saccharomyces cerevisiae*) on prepartum intake and postpartum intake and milk production of Jersey cows. J Dairy Sci 83:123–127

Dawson KA, Newman KE, Boling JA (1990) Effects of microbial supplements containing yeast and lactobacilli on roughage-fed ruminal microbial activities. J Anim Sci 68:3392–3398

Di Francia A, Masucci F, De Rosa G et al (2008) Effects of *Aspergillus oryzae* extract and a *Saccharomyces cerevisiae* fermentation product on intake, body weight gain and digestibility in buffalo calves. Anim Feed Sci Technol 140:67–77

Dias ALG, Freitas JA, Micai B et al (2018) Effect of supplemental yeast culture and dietary starch contents on rumen fermentation and digestion in dairy cows. J Dairy Sci 101:201–221

Elghandour MMY, Salem AZM, Castaneda JSM et al (2015) Direct-fed microbes: a tool for improving the utilization of low quality roughages in ruminants. J Integr Agric 14:526–533

Ferguson JD, Wu Z, Remsberg DW et al (2010) The influence of *Bacillus pumilus* 8G-134 on milk production of dairy cows in early lactation. J Dairy Sci 93(Suppl 1):871

Franklin ST, Newman MC, Newman KE et al (2005) Immune parameters of dry cows fed manna oligosaccharide and subsequent transfer of immunity to calves. J Dairy Sci 88:766–775

Frizzo LS, Sotto LP, Zbrun MV et al (2010) Lactic acid bacteria to improve growth performance in young calves fed milk replacer and spray-dried whey powder. Anim Feed Sci Technol 157:159–167

Galvão KN, Santos JE, Coscioni A et al (2005) Effect of feeding live yeast products to calves with failure of passive transfer on performance and patterns of antibiotic resistance in fecal *Escherichia coli*. Reprod Nutr Dev 45:427–440

Ganner A, Schatzmayr G (2012) Capability of yeast derivatives to adhere enter opathogenic bacteria and to modulate cells of the innate immune system. Appl Microbiol Biotechnol 95:289–297

Ghorbani GR, Morgavi DP, Beauchemin KA et al (2002) Effects of bacterial direct-fed microbials on ruminal fermentation, blood variables, and the microbial populations of feedlot cattle. J Anim Sci 80:1977–1985

Gomez-Basauri J, de Ondarza MB, Siciliano-Jones J (2001) Intake and milk production of dairy cows fed lactic acid bacteria and mannanoligosaccharide. J Dairy Sci 84:283–283

Hagg FM (2007) The effect of *Megasphaeraelsdenii*, a probiotic, on the productivity and health of Holstein cows. MSc dissertation. University of Pretoria

Harrison GA, Henken RW, Dawson KA et al (1988) Influence of addition of yeast culture supplement to the diets of lactating cows on ruminal fermentation and microbial populations. J Dairy Sci 71:2967–2975

Heinrichs AJ, Jones CM, Heinrichs BS (2003) Effects of mannanoligosaccharide or antibiotics in neonatal diets on health and growth of dairy calves. J Dairy Sci 86:4064–4069

Hristov AN, Varga G, Cassidy T et al (2010) Effect of *Saccharomyces cerevisiae* fermentation product on ruminal fermentation and nutrient utilization in dairy cows. J Dairy Sci 93:682–692

Ivan M, Neill L, Foster R et al (2000) Effects of *Isotricha*, *Dasytricha*, *Entodinium*, and total fauna on rumen fermentation and duodenal flow in weathers fed different diets. J Dairy Sci 83:776–787

Jensen GS, Hart AN, Schauss AG (2007) An anti-inflammatory immunogen from yeast culture induces activation and alters chemokine receptor expression on human natural killer cells and B lymphocytes in vitro. Nutr Res 27:327–335

Jensen GS, Patterson KM, Yoon I (2008) Nutritional yeast culture has specific anti-microbial properties without affecting healthy flora. Preliminary results. J Anim Feed Sci 17:247–252

Kenney NM, Vanzant ES, Harmon DL et al (2015) Direct-fed microbials containing lactate-producing bacteria influence ruminal fermentation but not lactate utilization in steers fed a high-concentrate diet. J Anim Sci 93:2336–2348

Khan RU, Naz S, Dhama K et al (2016) Direct-fed microbial: beneficial applications, modes of action and prospects as a safe tool for enhancing ruminant production and safeguarding health. J Pharmacol 12:220–231

Kim MK, Lee HG, Park JA (2011) Effect of feeding direct-fed microbial as an alternative to antibiotics for the prophylaxis of calf diarrhea in Holstein calves. Asian-Aust J Anim Sci 24:643–649

Krehbiel CR, Rust SR, Zhang G et al (2003) Bacterial direct-fed microbials in ruminant diets: performance response and mode of action. J Anim Sci 81:E120–E132

Kung Jr L (2001) Direct-fed microbials for dairy cows and enzymes for lactating dairy cows: new theories and applications. In: 2001 Pennsylvania State Dairy Cattle Nutrition Workshop, Grantville, pp 86–102

Kung L Jr (2006) Direct-fed microbial and enzyme feed additives. In: Lundeen T (ed) Direct-fed microbial, enzyme and forage additive compendium. Miller Publishing, Minnetonka

Kung L Jr, Hession AO (1995) Preventing in vitro lactate accumulation in ruminal fermentations by inoculation with *Megasphaeraelsdenii*. J Anim Sci 73:250–256

Lesmeister KE, Tozer PR, Heinrichs AJ (2004) Development and analysis of a rumen tissue sampling procedure. J Dairy Sci 87:1336–1344

Luan S, Duersteler M, Galbraith EA et al (2015) Effects of direct-fed *Bacillus pumilus*8G-134 on feed intake, milk yield, milk composition, feed conversion, and health condition of pre- and postpartum Holstein cows. J Dairy Sci 98:6423–6432

Lynch HA, Martin SA (2002) Effects of *Saccharomyces cerevisiae* culture and *Saccharomyces cerevisiae* live cells on in vitro mixed ruminal microorganism fermentation. J Dairy Sci 85:2603–2608

Magalhães VJA, Susca F, Lima FS et al (2008) Effect of feeding yeast culture on performance, health, and immunocompetence of dairy calves. J Dairy Sci 91:1497–1509

Mao HL, Mao HL, Wang JK et al (2013) Effects of *Saccharomyces cerevisiae* fermentation product on *in vitro* fermentation and microbial communities of low-quality forages and mixed diets. J Anim Sci 91:3291–3298

McAllister TA, Beauchemin KA, Alazzeh AY et al (2011) Review: the use of direct fed microbials to mitigate pathogens and enhance production in cattle. Can J Anim Sci 91:193–211

McDaniel MR (2009) The effects of dosing feedlot cattle with *Megasphaeraelsdenii* strain NCIMB 41125 prior to the introduction of a grain-rich diet. MSc thesis. Kansas State University

Miron J, Ben-Ghedalia D, Morrison M (2001) Invited review: adhesion mechanisms of rumen cellulolytic bacteria. J Dairy Sci 84:1294–1309

Moloney AP, Drennan MJ (1994) The influence of the basal diet on the effects of yeast culture on ruminal fermentation and digestibility in steers. Anim Feed Sci Technol 50:55–73

Murphy EA, Davis JM, Brown AS et al (2007) Oat β-glucan effects on neutrophil respiratory burst activity following exercise. Med Sci Sports Exerc 39:639–644

Mwenya B, Santoso B, Cbet S et al (2004) Effects of including β_1-4 galactooligosaccharides, lactic acid bacteria or yeast culture on methanogenesis as well as energy and nitrogen metabolism in sheep. Anim Feed Sci Technol 115:313–326

Mwenya B, Santoso B, Sar C et al (2005) Effects of yeast culture and galactooligosaccharides on ruminal fermentation in Holstein cows. J Dairy Sci 88:1404–1412

Newbold CJ, Wallace RJ, McIntosh FM (1996) Mode of action of the yeast *Saccharomyces cerevisiae* as a feed additive for ruminants. Br J Nutr 76:249–261

Newbold CJ, McIntosh FM, Wallace RJ (1998) Changes in the microbial population of a rumen-simulating fermenter in response to yeast culture. Can J Anim Sci 78:241–244

Newman KE (1994) Mannan-oligosaccharides: natural polymers with significant impact on the gastrointestinal microflora and the immune system. Alltech's. In: 10th Annual Symposium, Nottingham University Press, Nottingham

Nocek JE, Kautz WP (2006) Direct-fed microbial supplementation on ruminal digestion, health and performance of pre- and postpartum dairy cattle. J Dairy Sci 89:260–266

Nocek JE, Kautz WP, Leedle JAZ et al (2003) Direct fed microbial supplementation on the performance of dairy cattle during the transition period. J Dairy Sci 86:331–335

Nocek JE, Holt MG, Oppy J (2011) Effects of supplementation with yeast culture and enzymatically hydrolyzed yeast on performance of early lactation dairy cattle. J Dairy Sci 94:4046–4056

Nonnecke BJ, Kimura K, Goff JP (2003) Effects of the mammary gland on functional capacities of blood mononuclear leukocyte populations from periparturient cows. J Dairy Sci 86:2359–2368

Novak KN, Davis E, Wehnes CA et al (2012) Effect of supplementation with an electrolyte containing a *Bacillus*-based direct-fed microbial on immune development in dairy calves. Res Vet Sci 92:427–434

Oetzel GR, Emery KM, Kautz WP (2007) Direct-fed microbial supplementation and health and performance of pre- and postpartum dairy cattle: a field trial. J Dairy Sci 90:2058–2068
Ohtsuka HC, Watanabe M, Kohiruimaki T et al (2006) Comparison of two different nutritive conditions against the changes in peripheral blood mononuclear cells of periparturient dairy cows. J Vet Med Sci 68:1161–1166
Overton TR, Waldron MR (2004) Nutritional management of transition dairy cows: strategies to optimize metabolic health. J Dairy Sci 87(E Suppl):E105–E119
Pagnini C, Saeed R, Bamias G et al (2010) Probiotics promote gut health through stimulation of epithelial innate immunity. Proc Natl Acad Sci U S A 107:454–459
Pérez-Sotelo LS, Talavera-Rojas M, Monroy-Salazar HG et al (2005) In vitro evaluation of the binding capacity of *Saccharomyces cerevisiae* Sc47 to adhere to the wall of *Salmonella* spp. Rev Latinoam Microbiol 47:70–75
Pinos-Rodríguez JM, Robinson PH, Ortega ME et al (2008) Performance and rumen fermentation of dairy calves supplemented with *Saccharomycescerevisiae*[1077] or *Saccharomyces boulardii*[1079]. Anim Feed Sci Technol 140:223–232
Plata PF, Mendoza MGD, Barcena-Gamma JR et al (1994) Effect of a yeast culture (*Saccharomyces cerevisiae*) on neutral detergent fiber digestion in steers fed oat straw based diets. Anim Feed Sci Technol 49:203–210
Poppy GD, Rabiee AR, Lean IJ et al (2012) A meta-analysis of the effects of feeding yeast culture produced by anaerobic fermentation of *Saccharomyces cerevisiae* on milk production of lactating dairy cows. J Dairy Sci 95:6027–6041
Puniya AK, Salem AZM, Kumar S et al (2015) Role of live microbial feed supplements with reference to anaerobic fungi in ruminant productivity: a review. J Integr Agric 14:550–560
Qiao GH, Shan AS, Ma N et al (2010) Effect of supplemental *Bacillus* cultures on rumen fermentation and milk yield in Chinese Holstein cows. J Anim Physiol Anim Nutr 94:429–436
Ramin M, Alimon AR, Panandam JM et al (2008) Digestion of rice straw and oil palm fronds by microflora from rumen and termite bacteria, in vitro. Pak J Biol Sci 11:583–588
Reynolds CK, Aikman PC, Lupoli B et al (2003) Splanchnic metabolism of dairy cows during the transition from late gestation through early lactation. J Dairy Sci 86:1201–1217
Robinson PH, Garrett JE (1999) Effect of yeast culture (*Saccharomyces cerevisiae*) on adaptation of cows to postpartum diets and on lactational performance. J Anim Sci 77:988–999
Rose MD, Price B, Fink GR (1986) *Saccharomyces cerevisiae* nuclear fusion requires prior activation by alpha factor. Mol Cell Biol 6:3490–3497
Schingoethe DJ, Linke KN, Kalscheur KF et al (2004) Feed efficiency of mid-lactation dairy cows fed yeast culture during summer. J Dairy Sci 87:4178–4181
Seo JK, Kim SW, Kim MH et al (2010) Direct-fed microbials for ruminant animals. Asian-Aust J Anim Sci 23:1657–1667
Servin AL (2004) Antagonistic activities of lactobacilli and bifidobacteria against microbial pathogens. FEMS Microbiol Rev 28:405–440
Stein DR, Allen DT, Perry EB et al (2006) Effects of feeding propionibacteria to dairy cows on milk yield, milk components, and reproduction. J Dairy Sci 89:111–125
Sun P, Wang JQ, Zhang HT (2010) Effects of *Bacillus subtilisnatto* on performance and immune function of preweaning calves. J Dairy Sci 93:5851–5855
Tripathi MK, Karim SA (2010) Effect of individual and mixed live yeast culture feeding on growth performance, nutrient utilization and microbial crude protein synthesis in lambs. Anim Feed Sci Technol 155:163–171
Weidmeier RD, Arambel MJ, Walters JL (1987) Effect of yeast culture and *Aspergillus oryzae* fermentation extract on ruminal characteristics and nutrient digestibility. J Dairy Sci 70:2063–2068
Weiss WP, Wyatt DJ, McKelvey TR (2008) Effect of feeding propionibacteria on milk production by early lactation dairy cows. J Dairy Sci 91:646–652
Williams PE, Tait CA, Innes GM et al (1991) Effects of the inclusion of yeast culture (*Saccharomyces cerevisiae* plus growth medium) in the diet of dairy cows on milk yield and forage degradation and fermentation patterns in the rumen of steers. J Anim Sci 69:3016–3026

Wittum TE, Perino LJ (1995) Passive immune status at postpartum hour 24 and long-term health and performance of calves. Am J Vet Res 56:1149–1154

Yang WZ, Beauchemin KA, Vedres DD et al (2004) Effects of direct-fed microbial supplementation on ruminal acidosis, digestibility, and bacterial protein synthesis in continuous culture. Anim Feed Sci Technol 114:179–193

Yoon I, Garrett JE (1998) Yeast culture and processing effects on 24-hour *in situ* ruminal degradation of corn silage. In: Proceedings of world animal production conference

Yoon IK, Stern MD (1996) Effects of *Saccharomyces cerevisiae* and *Aspergillus oryzae* cultures on ruminal fermentation in dairy cows. J Dairy Sci 79:411–417

Yoon IK, Garrett JE, Cox DJ (1997) Effect of yeast culture supplementation toalfalfa-grass hay diet on microbial fermentation in continuous culture of rumen contents. J Anim Sci 75(Suppl 1):98

Yuan K, Mendonça LGD, Hulbert LE et al (2015) Yeast product supplementation modulated humoral and mucosal immunity and uterine inflammatory signals in transition dairy cows. J Dairy Sci 98:3236–3246

Zaworski EM, Shriver-Munsch CM, Fadden NA et al (2014) Effects of feeding various dosages of *Saccharomyces cerevisiae* fermentation product in transition dairy cows. J Dairy Sci 97:3081–3098

Zhu W, Zhang BX, Yao KY et al (2016) Effects of supplemental levels of *saccharomyces cerevisiae* fermentation product on lactation performance in dairy cows under heat stress. Asian Aust J Anim Sci 29:801–806

Zhu W, Wei ZH, Xu NN et al (2017) Effects of *Saccharomyces cerevisiae* fermentation products on performance and rumen fermentation and microbiota in dairy cows fed a diet containing low quality forage. J Anim Sci Biotech 8(3):36

Effects of Exogenous Enzymes on the Nutritive Value of Some Fibrous Forage in Ruminant

**Abdelfattah Zeidan Mohamed Salem,
Mona Mohamed Mohamed Yasseen Elghandour, Moyosore Joseph Adegbeye,
Javier Hernández Meléndez, José Luis Ponce-Covarrubias,
and Pedro Enrique Hernández Ruiz**

Abstract Ruminant feeding efficiency will always be an important research topic. Thus there is a need to enhance the digestibility of poor-quality fibrous forages due to their lignified cell walls that restrict nutrient availability. A potentially cost-effective means of improving nutrient availability is by supplementation with exogenous enzymes. This chapter explores the influence of exogenous enzymes added to fibrous forages on the nutritive value of diets for ruminants. The sources of exogenous enzymes and their role in improving the utilization of animal feeds are discussed. Moreover, the effects of exogenous enzyme supplementation on ruminal degradation, microbial protein synthesis, and nutrient digestibility were reviewed, together with their effects on animal growth performance, milk yield and the quality of milk and meat.

A. Z. M. Salem (✉) · M. M. M. Y. Elghandour
Facultad de Medicina Veterinaria y Zootecnia, Universidad Autónoma del Estado de México, Toluca, Estado de México, Mexico

M. J. Adegbeye
Department of Animal Production and Health, Federal University of Technology, Akure, Nigeria

J. H. Meléndez
Facultad de Ingeniería y Ciencias, Universidad Autónoma de Tamaulipas, Centro Universitario "Adolfo López Mateos" Cd Victoria Tamaulipas, Mexico

J. L. Ponce-Covarrubias · P. E. H. Ruiz
Escuela Superior de Medicina Veterinaria y Zootecnia No. 3, Universidad Autónoma de Guerrero, Tecpan de Galeana, Mexico

A. Z. M. Salem et al. (eds.), *Exogenous Enzymes as Feed Additives in Ruminants*, https://doi.org/10.1007/978-3-031-27993-5_3

1 Introduction

Many local plant species are undervalued as forages for ruminants, because of inadequate knowledge on their nutritive value, especially the fiber content and level of antinutritional factors, which can limit feed efficiency and overall animal performance. The proficiency with which ruminants acquire energy from diets comprising complex plant polysaccharides has been an enduring important research question. Improvements in the efficiency of feed use may enable ruminants to produce more and better-quality meat and milk, thereby contributing to meeting the ever-increasing demands of food for the increasing human population. Considering the economy of food production from animals and their physiological requirements to maintain rumen health, forage constitutes the major bulk of ruminant diets (Krause et al. 2003). However, the conditions within the rumen for effective fiber digestion are often suboptimal, leading to poor digestibility of forage cell walls, which in turn limits nutrient availability to the animal.

The use of exogenous enzymes is regarded as an inexpensive method of enhancing feed conversion efficiency (Krause et al. 2003). There has been a considerable amount of published research on beneficial effects of using exogenous enzymes to improve feed efficiency, nutrition, and productivity in ruminants (Krause et al. 2003; Beauchemin and Holtshausen 2010). The application of fibrolytic enzymes in ruminants feed often results in boosting fiber digestion and improving intake of feed in vivo. Also, they have been reported to increase yield and better composition of milk (Gado et al. 2009).

Exogenous fibrolytic enzyme treatment resulted in increased milk production efficiency in early lactation of dairy cows (Holtshausen et al. 2011). Arriola et al. (2011) ascribed the improvement in feed efficiency to a rise in the total gastrointestinal tract digestibility without a change in ruminal degradability. Holtshausen et al. (2009) ascribed the increase in milk production efficiency to an increase in fiber digestibility (based on in vitro assay) of dietary forages. However, responses to ruminal fermentation were not investigated in that study. It is apparent that exogenous enzymes added to fibrous, poor-quality forages, or crop residues can be practiced in animal feeding without compromising the energy intake and milk production.

A prepared exogenous enzyme mixture of polysaccharidases (ZADO®), obtained from anaerobic bacteria, was reported to increase nutrient digestibility, N balance and ruminal fermentation, and milk yield of cows receiving diets based upon crop residues and by-products (Gado et al. 2009). Salem et al. (2011) found higher feed efficiency, average daily gain, in goat and sheep fed on wheat straw.

The purpose of this chapter is to elucidate the benefit of using exogenous enzymes on some fibrous feed ingredients fed to ruminants.

2 Definition of Exogenous Enzymes

Exogenous enzymes are enzyme products that can be used to supplement animal feeds to enhance the digestion process. Examples of exogenous enzymes include cellulases (hydrolyzing cellulose), proteases (hydrolyzing proteins), and amylase (hydrolyzing starch). Recent advances in biotechnology, which makes enzyme production cost-effective, have led to a wide range of enzyme products becoming available. With improvements in knowledge of rumen functioning and techniques for evaluating enzyme activity, interest in the use of exogenous enzyme products in ruminant production systems has increased (Sujani and Seresinhe 2015). Exogenous enzymes used in ruminant diets can be mainly categorized as proteolytic, amylolytic, and fibrolytic enzymes, depending on the substrate on which the enzyme acts (Sujani and Seresinhe 2015).

The aim of supplementing ruminant diets with exogenous enzymes is to improve the digestibility of the dietary plant cell walls, and consequently improve the efficiency of feed utilization. Having knowledge of the complexity of the rumen ecosystem and understanding the nature of interactions between rumen microbes and plant cell walls is key for efficient use of exogenous enzymes in enhancing the utilization of feeds by ruminant (Meale et al. 2014).

The use of in vitro methods to screen exogenous enzymes offers a fast and inexpensive way to evaluate a large number of different products with various substrates and could be applied for prediction of possible in vivo responses (Jalilvand et al. 2008). However, in vivo animal trials measuring production responses to different known doses of exogenous enzymes represent the final proof of the beneficial effects of enzyme supplementation of animal feeds.

3 Sources of Exogenous Enzymes

The exogenous enzymes used to enhance the nutritive value of ruminant feeds are mostly sourced from microbial cultures involving four bacterial species and three fungal species, in addition to some yeasts. The main processes used to produce the enzymes are submerged fermentation, in which the substrates and other nutrients are in liquid form, and solid-state fermentation, which use solid substrates associated with low levels of moisture being present; several other biotechnological methods are also involved (Sujani and Seresinhe 2015). The exogenous fibrolytic enzymes are classified according to the specific substrates upon which they act. The enzyme cellulase hydrolyzes the plant cell wall fiber to cello-oligosaccharides, cellobiose, or glucose with a collective activity of three enzymes, including exoglucanase, β-glucosidase, and endoglucanase (Zhang and Lynd 2004). Xylanase hydrolyses 1,4-β-D-xylosidic linkages in xylans found in the structural component of plant cell walls, generally known as hemicellulose. The group of xylanolytic enzymes includes endo-β-1,4-xylanase (EC 3.2.1.8, 1,4-β-D xylanxylanohydrolase) that breaks the main chain of xylans, β-D-xylosidase (EC 3.2.1.37, 1,4-β-xylanxylohydrolase) that hydrolyzes xylo-oligosaccharides to D-xylose, as well as a selection of debranching enzymes such as acetyl esterases, α-glucuronidases, and α-L-arabino-furanosidases (Kamble and Jadhav 2012).

Most enzymes used in the livestock industry are obtained from the bacteria, *Enterococcus faecium* spp., *Lactobacillus plantarum*, *L. acidophilus*, and *Bacillus subtilis* and the fungi, *Trichoderma reesei* and *Aspergillus oryzae* (McAllister et al. 2001). The enzymes are produced by microbial fermentation that starts with a growth media and a seed culture. The activity and types of enzymes generated vary based on the conditions and substrates used for the culture and the selected strain of the microbes (Lee et al. 1998).

A wide range of enzymes are required for the degradation of complex structural carbohydrates that comprise plant cell walls (Morgavi et al. 2013). Although conventional enzymes are mainly xylanases or cellulases, secondary enzymes, such as proteases, esterases, pectinases, and amylases, are normally present, since the preparations rarely comprise only a single pure enzyme (McAllister et al. 2001). This diversity is beneficial since several substrates could be targeted with a single product. Degradation of hemicellulose and cellulose alone needs several glycosidic hydrolases (Krause et al. 2003) and alterations in the activity of specific enzymes and their relative proportions affect the overall efficiency of degradation of the forage cell wall (McAllister et al. 2001). Research results indicate that a mixture of enzymes developed for enhancing the digestibility of a particular diet is better than targeting a specific enzyme for a particular feed (Beauchemin et al. 2003). The inclusion of exogenous enzymes in feeds without considering the relevant substrates could lead to a variation in results. Highly variable results of enzyme supplementation may discourage farmers from using enzymes in ruminant feeds (Table 1).

Table 1 Summary of the effects of adding exogenous enzymes to ruminant feeds

Enzyme	Dose	Impact	Animal species	References
ZADO®[a]	10 g/h/d	OM[b] and DM digestibility and live weight	Growing lamb	Valdes et al. (2015)
ZADO®	40 g/h/d	ADF and NDF digestibility	Cows	Gado et al. (2009)
ZADO®	Not specified	Fiber digestibility	Ossimi male lambs	Gado et al. (2011)
Celluase plus®	2 mL cellulase per kg DMI	DM and NDF	Alpine goats	Rojo et al. (2015)
Cellulases	Not specified	OM, ADE and NDF digestibility	Pelibuey lambs	López-Aguirre et al. (2016a)
Xylanase and cellulase	0, 1, 3 μL per 0.5 g DM	OM digestibility	In vitro	López-Aguirre et al. (2016b)
ZADO®	0, 6, 12, 24 mg/g DM	OM digestibility	In vitro	Elghandour et al. (2013)
Xylanase and cellulase	2 g/h/d	Nutrient digestion and milk yield	Cows	Yang et al. (1999)
ZADO®	10 g/h/d	Growth performance	Suffolk lambs	Rivero et al. (2012)

[a]ZADO®: a commercial product comprising an enzyme mixture of cellulase, xylanase, protease, and α-amylase
[b]*OM* organic matter, *DM* dry matter, *ADF* acid detergent fiber, *NDF* neutral detergent fiber, *DMI* dry matter intake
Cellulase Plus®: a commercial cellulase product

4 Roles of Exogenous Enzymes in Improving Nutritive Value of Animal Feeds

The use of exogenous enzymes as a feed additive has been shown to enhance the nutritive value of tree foliage in animal feeds due to an increase in the degradation of the dietary fiber (Salem et al. 2015a). Salem et al. (2015a) reported that the application of exogenous enzymes can improve the nutritive value of fodder trees used as supplements in ruminant diet. The inclusion of exogenous enzymes can enhance the number of bacteria, fiber digestibility, and capabilities of rumen bacteria for degradation of feed consumed by the animal. The supplementary enzymes may also help to increase the availability of protein for microbial metabolism, thereby increasing the amount of metabolizable energy and fiber digestibility in the diet (Polderman et al. 1955; Salem et al. 2012). The exogenous fibrolytic enzymes can work synergistically with endogenous enzymes produced by the rumen microbes with a resultant increase in the nutritive value and digestion of fibrous feeds (Morgavi et al. 2001).

Dairy cows fed forages supplemented with exogenous enzymes had more dry matter intake and produced 5–25% more milk than their counterparts receiving no enzyme supplements (Gado et al. 2009). In feedlot cattle, exogenous fibrolytic enzyme was successfully used to boost live weight gain by up to 16% (Salem et al. 2013). In lactating goats, exogenous enzyme supplementation resulted in an increased yield of milk with a lower fat content (Stella et al. 2007). Titi and Lubbadeh (2004) reported that exogenous enzyme supplementation increased milk yield and milk fat and protein levels in lactating sheep, but in goats, enzyme supplementation increased milk yield, but milk composition was not affected.

Application of exogenous enzymes phytases and nonstarch polysaccharidases is very common in poultry diets and frequently used in swine rations. The bulk of research carried out using these enzyme products in diets for poultry and pigs has focused on the determination of their influence on nutrient digestibility and/or animal performance (Bedford and Cowieson 2012).

5 Impact of Exogenous Enzymes

5.1 Fermentation of Fibrous Feeds

Considerable variation in responses to exogenous enzyme supplementation of feeds on ruminal fermentation has been reported in literature. Many studies have reported increases in diet digestibility after treatment of feeds with exogenous enzymes. Arriola et al. (2011) observed a decrease in the acetate: propionate ratio in rumen fluid and an increase in total volatile fatty acid concentration. In contrast, increases in the proportion of acetate in rumen fluid following exogenous fibrolytic enzyme supplementation have been reported by Gado et al. (2009) and Beauchemin et al.

(2000). However, Beauchemin et al. (1999) and Yang et al. (1999) found no effect of exogenous fibrolytic enzymes on ruminal fermentation.

Pretreatment of total mixed ration with exogenous enzymes may help increase the availability of nutrients to ruminal microbes, due to increased colonization and attachment of microbes to feed particles (Wang et al. 2001). *Fibrobacter succinogenes*, which is a fermentation product from primary digesters, can attract *Ruminobacter amylophilus* and *Selenomonas ruminantium* as secondary digesters to attach to and colonize feed particles (Cheng and McAllister 1997). McGinn et al. (2004) reported that supplementation with a proteolytic enzyme obtained from yeast had no effect on enteric CH_4 emissions from cattle receiving a diet based on barley silage and barley grain. Arriola et al. (2011) examined the effects of exogenous fibrolytic enzyme supplementation of high and low concentrate rations fed to dairy cows. Using a fermentation balance approach to estimate methane production from volatile fatty acid data, a reduction in methane production was observed in cows receiving the high-concentrate diet. Results of in vitro studies have shown that enhancement in fiber degradation by exogenous fibrolytic enzymes was often accompanied by a decrease in acetate: propionate ratio in the fermentation fluid (Eun and Beauchemin 2005; Eun et al. 2007). Beauchemin et al. (2008) suggested that exogenous enzyme feed supplements could assist in reducing emissions of enteric CH_4. However, increased enteric CH_4 production with an increase in enzyme dosage reported by Chung et al. (2012) reinforces the need for in vivo evaluation with direct measurement of CH_4 emissions. An increase in CH_4 production with increasing enzyme dosage implies that more energy was lost in form of CH_4 during ruminal fermentation. This suggests that if enzyme supplementation increases ruminal nutrient availability, methane production would be increased (Zhou et al. 2011). The increase in feed efficiency is perhaps due to a change in the microbial communities in response to enzyme supplementation (Holtshausen et al. 2011). Using the same enzyme preparation and diet, Zhou et al. (2011) observed that enzyme supplementation had no effect on the overall microbe population size, although the relative activities of individual methanogen species was altered. A detailed study is required to investigate the long-term effect of exogenous fibrolytic enzyme on rumen fermentation characteristics and enteric CH_4 production.

5.2 *Nutrient Digestibility*

The incorporation of a high proportion of grain into ruminant diets decreases fiber digestion because of poor fibrolytic enzyme activity in the rumen, caused by seeming preference of rumen microbes for different feed structures like nonstructural carbohydrates and depression in rumen pH (Nozière et al. 1996; Faniyi et al. 2019). This may bring about conditions in which addition of exogenous enzymes could be advantageous (Vallejo et al. 2016). Rumen microbial enzymes can work synergistically with exogenous fibrolytic enzymes to increase the digestibility and nutritional value of fibrous feeds (Morgavi et al. 2001).

Following exogenous enzyme supplementation, Arriola et al. (2011) associated the enhanced feed utilization efficiency with increased total-tract digestion, since ruminal degradability, measured in sacco, was not affected. Gado et al. (2009) investigated the effect of the multienzyme preparation, ZADO®, added to a total mixed ration of dairy cows on nutrient digestion. They reported an 8–16% higher digestibility of acid detergent fiber, neutral detergent fiber, dry matter, and organic matter compared with values obtained from those receiving the control treatment (no enzyme added). Table 2 shows data from Gado et al. (2009), indicating the influence of enzyme supplementation on intake, digestibility, nitrogen balance and ruminal characteristics.

Using Ossimi male lambs, Gado et al. (2011) investigated the effect on growth performance and digestion of ensiled orange pulp (EOP) supplementation of feed with or without ZADO® exogenous enzyme product. They reported lower nutrient digestion in the EOP treatment compared with the EOP plus ZADO® treatment. Digestibility of dry matter, acid detergent fiber (ADF), and neutral detergent fiber (NDF) was increased by 18, 47, and 93%, respectively, in lambs fed with EOP plus ZADO®. The feed conversion efficiency was increased by 31% and 19% in lambs on the EOP plus ZADO® and EOP treatments, respectively, compared with the lambs on the control (unsupplemented) treatment. Compared with the lambs on the control treatment, liveweight gain increased by 54% in the EOP lambs and by 92% in EOP plus ZADO® lambs. Nevertheless, dry matter intake was low in all treatment groups.

Gado et al. (2009) studied the effects of supplementing a total mixed ration based on maize silage and cereal concentrate with exogenous fibrolytic enzymes on milk yield and composition in dairy cows. Enzyme supplementation generally induced changes in ruminal volatile fatty acid concentrations. The total short chain fatty acid concentrations were increased by 7.4%, acetate concentration by 6.7% and propionate by 13.7%. Consequently, the acetate: propionate ratio was reduced by 6.5%. In addition, enzyme supplementation increased ruminal microbial nitrogen synthesis. Using Alpine dairy goats, Rojo et al. (2015) reported that supplementing diets with an exogenous cellulase preparation (Celluase Plus®) increased dry matter intake and digestibility of dry matter, neutral detergent fiber and organic matter compared with the unsupplemented diet. López-Aguirre et al. (2016a, b) investigated the effect of exogenous cellulase (Cellulase Plus®) and its mode of administration on nutrient digestibility in Pelibuey lambs. They reported that digestibility of acid detergent fiber, neutral detergent fiber, organic matter, and dry matter when the enzyme was administered either by incorporation in the feed or by direct oral dosage; the digestibility of crude protein was not affected. However, feed conversion ratio and digestibility of acid detergent fiber were improved only when the enzyme was incorporated in the diet. Thus, the influence of exogenous enzyme addition is dependent to some degree on the method by which the enzyme is administered. Data from previous studies reported by Valdes et al. (2015), Rojo et al. (2015), and Salem et al. (2013) are summarized in Table 3.

Table 2 Feed intake, digestibility, nitrogen balance, and rumen fluid characteristics in dairy cows fed with a total mixed ration (control treatment) and the total mixed ration supplemented with a mixed enzyme preparation (ZADO®) - (Gado et al. 2009)

	Control	ZADO®[a]	SEM	P value
Intake (kg/d)				
Dry matter	16.1	18.2	0.21	0.049
Organic matter	14.1	16.4	0.14	0.048
Neutral detergent fiber	7.1	7.4	0.23	0.192
Acid detergent fiber	4.04	4.57	0.111	0.087
Digestibility (g/kg)				
Dry matter	663	743	2.1	0.045
Organic matter	667	741	2.9	0.047
Neutral detergent fiber	418	584	2.8	0.049
Acid detergent fiber	401	532	2.3	0.046
Nitrogen balance				
N intake (g/d)	1654	1870	54.1	0.07
Urinary N (g/d)	509	542	12.6	0.11
Fecal N (g/d)	659	711	14.3	0.17
N balance (g/d)	468	617	11.9	0.06
Rumen fluid characteristics				
pH	6.1	5.9	0.24	0.41
Before feeding (0 h)				
Total short chain fatty acids (mmol/L)	111	122	2.1	0.34
Individual SCFA (molar %)				
Acetate (A)	61.0	64.8	1.30	0.05
Propionate (P)	17.8	18.1	0.83	0.13
Butyrate	11.3	11.9	0.81	0.24
Acetate: Propionate ratio	3.43	3.85	1.162	0.14
Ammonia N (mg/L)	55	67	0.37	0.04
Post feeding (3 h)				
Total short chain fatty acids (mmol/L)	119.2	128	3.6	0.04
Individual SCFA (molar %)				
Acetate (A)	60.0	64.0	1.2	0.04
Propionate (P)	18.3	20.8	0.87	0.01
Butyrate	10.9	11.0	0.96	0.31
Acetate: Propionate ratio	3.28	3.08	0.071	0.01
Ammonia N (mg/L)	110	126	2.3	0.05
Microbial N synthesis (g/d)	190	220	9.6	0.04
Urinary purine derivative excretion				
Uric acid (mmol/d)	22.4	24.6	0.67	0.16
Allantoin (mmol/d)	308	304	10.4	0.26

[a]ZADO®: A commercial product obtained from *Ruminococcus flavefaciens* comprising cellulase, xylanase, protease, and α-amylase enzymes

Table 3 Data from Valdes et al. (2015), Rojo et al. (2015), and Salem et al. (2013) summarizing the impact of exogenous enzymes on feed intake and nutrient digestibility in various livestock species

	Treatment					
Parameter[a]	Control	Enzyme[b]	*Salix babylonica*[c]	Enzyme plus *S. babylonica*	SEM	P value
	No plant extract	No plant extract	Plant extract 30 mL/h/day	Plant extract 30 mL/h/d		
Valdes et al. (2015)	No enzyme	ZADO®	No enzyme	ZADO®		
Suffolk lambs		10 g/h/d		10 g/h/d		
DM intake (g/d)	567	636	574	699	21.4	<0.001
OM intake (g/d)	521	578	528	685	15.9	0.01
CP intake (g/d)	412	465	437	452	14.3	0.002
NDF intake (g/d)	160	228	186	201	13.6	0.001
ADF intake (g/d)	108	154	125	135	10.4	<0.001
DM digestibility						
OM digestibility	0.746	0.878	0.810	0.889	0.0316	0.001
CP digestibility	0.721	0.724	0.727	0.722	0.0245	0.001
NDF digestibility	0.601	0.665	0.676	0.694	0.1625	0.001
ADF digestibility	0.455	0.539	0.592	0.561	0.0943	0.001
Rojo et al. (2015)		Cellulase plus®				
Alpine dairy goats		2 ml/kg DM				
DM intake (g/d)	3.1	3.5	–	–	0.12	0.049
OM intake (g/d)	2.9	3.2	–	–	0.11	0.049
NDF intake (g/d)	1.11	1.23	–	–	0.041	0.049
OM digestibility	0.76	0.80	–	–	0.013	0.001
DM digestibility	0.74	0.77	–	–	0.011	0.001
NDF digestibility	0.62	0.68	–	–	0.011	<0.001

(continued)

Table 3 (continued)

	Treatment					
Parameter[a]	Control	Enzyme[b]	*Salix babylonica*[c]	Enzyme plus *S. babylonica*	SEM	P value
Salem et al. (2013)		ZADO®				
Beef cattle		40 g/h/d				
DM intake (kg/d)	7.3	7.8	–	–	0.3	0.11
DM digestibility	617	691	–	–	12.2	0.04
OM digestibility	674	753	–	–	151	0.01
CP digestibility	835	874	–	–	7.6	0.04
NDF digestibility	417	508	–	–	12.3	0.01
ADF digestibility	322	408	–	–	15.1	0.01

[a]*DM* dry matter, *OM* organic matter, *NDF* neutral detergent fiber, *ADF* acid detergent fiber, *CP* crude protein

[b]ZADO®: A commercial product obtained from *Ruminococcus flavefaciens* comprising cellulase, xylanase, Protease, and α-amylase enzymes; Cellulase Plus®: A commercial product obtained from *Trichoderma longibrachiatum*, containing cellulase and β-glucanase

[c]Extract of the willow species *Salix babylonica*, obtained by macerating leaves in methanol/ethanol/water

5.3 Ruminal Degradability and Gas Production

Generally, fiber degradation in rumen is not fully efficient as indicated by the presence of fermentable fiber in feces (Krause et al. 2003). Mendoza et al. (2014) suggested that feed supplementation with exogenous fibrolytic enzymes may be effective in acting on the potentially digestible fraction of dietary fiber. Enzymes added to feeds are effective in increasing fiber degradation when structural barriers that interrupt microbial colonization of digestible fractions in the consumed plants are eliminated (López-Aguirre et al. 2016a). Gado et al. (2013) investigated the effect of ensiling rice straw, dried distillers' grains and a 9:1 mixture of the two feedstuffs, for 30 days with a liquid exogenous enzyme product containing cellulases, xylanases, and α-amylase (ZAD®), on levels of fiber fractions and on in vitro ruminal degradability. Treatment with the exogenous enzyme product enhanced degradation of acid detergent fiber, dry matter, and neutral detergent fiber. The feed degradation fractions (rapid soluble fraction and insoluble, but potentially degradable fraction) and degradation rate were increased when the enzyme inclusion rate was 3 L/tonne, except for rice straw where the degradation rate values for acid detergent fiber and neutral detergent fiber were not influenced by the exogenous enzyme treatment.

Togtokhbayar et al. (2015) investigated the effect of xylanase on in sacco degradation of wheat straw and showed that the dry matter degradation decreased linearly with increasing enzyme dosage. Although the in sacco neutral detergent fiber degradation remained unchanged with xylanase addition, an intermediate xylanase dosage level enhanced rumen degradability. Salem et al. (2015b) reported positive effects of glucoamylase extracts on in vitro degradability of two total mixed rations that contained 25% sorghum grains and 25% maize. For both diets, supplementation with glucoamylase had no effect on the digestibility of acid detergent fiber, neutral detergent fiber, and dry matter. Regardless of the level of exogenous enzyme supply, the pattern of degradation of sorghum was generally lower than that of maize.

López-Aguirre et al. (2016a, b) studied the effect of three exogenous fibrolytic enzyme products (xylanase, cellulase and a 1:1 blend of xylanase and cellulase) at varying dosages (0, 1, 3 μL/0.5 g DM) on the in vitro fermentation of feeds formulated for growing lambs. Bottles containing the samples were incubated at 39 °C for 96 h. Data showed that the use of cellulases improved degradation of dry matter and acid detergent fiber and increased the asymptotic gas production value; interactions between the enzyme dosage and products were significant.

5.4 *Nitrogen Balance and Microbial Protein Synthesis*

Most tropical forages and fibrous crop residues have a poor digestibility and low protein content (Salem et al. 2015c). Microbial protein is a vital component of the nitrogen flow after foregut fermentation in ruminants (Valdes et al. 2015). Supplementation of animal diets with exogenous enzymes has been shown to improve the nutritional value of tree foliage owing to the actions of enzymes in enhancing the degradation of dietary fiber (Salem et al. 2015a). Wang et al. (2001) investigated the effects of exogenous enzyme by different modes of application on the synthesis of microbial protein by using the rumen simulation technique (Rusitec) and reported an increased microbial protein synthesis (assessed by ^{15}N incorporation) when crude xylanase was added to a diet 24 h before fermentation began.

5.5 *Milk Production, Composition and Quality*

There are several published reports claiming positive effects of dietary supplementation with exogenous enzymes on the composition of cow milk. Beauchemin et al. (2000) reported a 2% increase in protein content of milk after feed supplementation with an enzyme product containing endoglucanase, β-glucanase, and xylanase. There are other reports of an increase in milk protein or fat content (Bowman et al. 2002; Sutton et al. 2003; Eun and Beauchemin 2005). There are also some published cases of improvements in milk yield because of enzyme supplementation. Yang et al. (1999) observed an additional milk yield of 1.9 kg/d when an exogenous enzyme product, comprising predominantly of xylanase and cellulase, was used to supplement hay at a relatively high enzyme inclusion rate (2 g/kg hay). The result was interpreted as being due to a 4% increase in total tract organic matter digestibility. However, there are different research reports showing conflicting responses to enzyme supplementation of feed for dairy cows. Some studies reported no effect of enzyme treatment of diets on milk yield (DeFrain et al. 2005; Bernard et al. 2010; Peters et al. 2010; Ferraretto et al. 2011) or milk composition (Bernard et al. 2010; Arriola et al. 2011). Holtshausen et al. (2011) investigated the effectiveness of an exogenous enzyme preparation (Econase RDE), containing endoglucanase and xylanase as a supplement to dairy cow diets. Enzyme supplementation of diets fed to lactating Holstein cattle increased fat-corrected milk yield in a dosage-dependent manner. Holtshausen et al. (2011) tentatively attributed these results to an improvement in fiber digestibility of dietary forages, since, in the same study, in vitro measurements indicated that enzyme supplementation increased fiber degradation in one of the component dietary forages (alfalfa hay), but fiber degradation measured using the in sacco method was unaffected. Likewise, Arriola et al. (2011) observed no effects of exogenous enzyme supplementation for high and low concentrate cattle diets in terms of ruminal fiber degradation measured in sacco, despite increases in total-tract digestibility and in efficiency of milk production. The enzyme preparation investigated in the study by Arriola et al. (2011), which contained cellulose, xylanase, and esterase, was obtained from a strain of *Trichoderma longibrachiatum*. Diverse published data on production responses to exogenous enzyme supplementation suggest that it is important for the type and dosage of enzyme preparations to be standardized before adding them into the diets for dairy cattle.

The adoption of in vitro screening techniques could aid determination of appropriate application rates for exogenous enzyme preparations and prediction of production responses when used to supplement specific ruminant feeds. However, it is not easy to evaluate the consistency of animal responses to each enzyme product, as most of the enzymes preparations are at experimental stage and their activities could vary over time. Gado et al. (2009) investigated the influence of the commercial fibrolytic mixed enzyme product, ZADO® addition to a total mixed ration, based on maize silage and concentrate fed to dairy cows, on the production and composition of milk. Milk yield and milk energy were greater for cows fed with the diet supplemented with the exogenous enzyme product (Table 4). However, exogenous enzyme supplementation had no effect on the fat, protein, or lactose concentrations in milk, although daily yield of milk protein was increased.

Rojo et al. (2015) examined the effects of cellulase addition to a diet fed to French Alpine goats on the production and composition of their milk. Goats receiving the cellulase supplement yielded more milk with higher energy density, and higher concentrations of fat, protein, and lactose compared with that of goats receiving the unsupplemented ration. Therefore, milk energy yield, energy-corrected yield, milk energy, and total solids contents were also increased (Table 5).

Milk fat from the goats fed with the cellulase-supplemented diet had higher concentrations of mono-unsaturated acids such as palmitoleic acid (C16:1) and cis-10-heptadecanoic acid (C17:1), with lower levels of saturated fatty acids, and the polyunsaturated fatty acids, linolenic acid (C18:2), and linoleic acid (C18:3).

Table 4 Milk production and composition from dairy cows fed with a total mixed ration with or without a supplement of a commercial mixed enzyme product (ZADO®) - (Gado et al. 2009)

	Treatment[a]			
	TMR[b]	TMR + ZADO®[c]	SEM	P value
Milk yield				
Total production (kg/d)	12.8	15.7	0.85	0.046
Fat (kg/d)	0.50	0.60	0.041	0.126
Protein (kg/d)	0.45	0.57	0.022	0.047
Lactose (kg/d)	0.58	0.71	0.054	0.141
Total energy (MJ/d)	58.2	70.7	8.15	0.041
Milk composition (g/kg)				
Fat	39	38	2.4	0.163
Protein	35	36	1.7	0.242
Lactose	45	45	2.2	0.361

[a]Treatment: Ten animals per treatment group
[b]*TMR* Total mixed ration, based on maize silage (70%, dry matter basis) and concentrate (30%, dry matter basis); concentrate contained mainly ground maize grain and cottonseed cake
[c]ZADO®: A commercial product obtained from *Ruminococcus flavefaciens* comprising cellulase, xylanase, protease, and α-amylase enzymes at a rate of 40 g/h/d

5.6 Growth Performance and Meat Quality

Several studies have shown that the use of exogenous enzyme preparations added to diets in varying compositions and doses increased the dry matter intake and growth of a range of young ruminants, including beef steers, goat kids, and lambs (McAllister et al. 1999; Titi 2003; Cruywagen and Van Zyl 2008). Conversely, other published studies have reported no effects of exogenous enzyme supplementation on intake or growth of beef steers or lambs (Mora-Jaimes et al. 2002; Miller et al. 2008a; b; Awawdeh and Obeidat 2011).

Rivero et al. (2012) investigated the effect of daily addition to a maize silage/maize-based concentrate diet, of a mixture of 10 g of the exogenous enzyme preparation, ZADO®, and 30 mL of an extract from *Salix babylonica* leaves on the growth of Suffolk lambs. Results showed that *S. babylonica* and ZADO® individually or in combination in feeds did not affect growth or dietary intake. Blood chemistry, cellular immune response, and animal health were unaffected by the plant extract and the enzyme treatments. Even though 59 compounds were found in *S. babylonica*, none of these compounds were examined individually for any effects on animal nutrition.

However, several studies have shown that the addition of crude *S. babylonica* extracts to feeds, without exogenous enzyme supplementation, resulted in enhanced nutrient digestion and animal growth in lambs (Salem et al. 2011, 2014). Such differences in results among separate studies could perhaps be attributed to variation in plant secondary metabolites from *S. babylonica* extracts and their impact on the activity of ruminal microorganisms (Salem et al. 2012, 2014; Jiménez-Peralta et al. 2011).

Table 5 Milk production and composition during the first third of the lactation period (60 days) from French Alpine dairy goats fed with a total mixed ration with or without a supplement of a cellulase enzyme product (Celluase Plus®) - (Rojo et al. 2015)

	Treatment[a]			
	TMR[b]	TMR + Cellulase Plus®[c]	SEM	P value
Milk yield				
Total production (kg/d)	3.1	3.4	0.13	0.042
Energy-corrected milk (ECM, kg/d)	2.6	3.3	0.12	0.001
Milk energy output (MJ/day)	8.3	10.5	0.37	0.001
Total solids (kg/d)	315	392	13.3	0.001
Fat (kg/d)	103	136	5.9	0.001
Protein (kg/d)	87	105	3.4	0.001
Lactose (kg/d)	125	151	5.1	0.002
Milk composition				
Milk energy content (MJ/kg)	2.8	3.1	0.18	0.008
Total solids (g/kg)	104	114	2.6	0.008
Fat (g/kg)	34	40	1.5	0.016
Protein (g/kg)	29	31	1.4	0.003
Lactose (g/kg)	41	44	2.0	0.050
Density (g/mL)	1.026	1.028	0.0012	0.003
Feed efficiency				
Milk (milk yield/DMI[d])	0.99	1.01	0.053	0.772
Energy corrected milk (ECM/DMI)	0.84	0.96	0.044	0.066

[a]Treatment: Twelve animals per treatment group
[b]*TMR* Total mixed ration, based on alfalfa hay, sorghum grain, and maize grain
[c]Cellulase Plus®: Commercial cellulase preparation added to the TMR (2 mL/kg dry matter)
[d]*DMI* Dry matter intake (kg/d)

Buendía et al. (2014) reported that the intake and growth performance of lambs was not affected by the addition of phytase to the diet, although diet digestibility was enhanced. The application of exogenous phytase also reduced the fecal excretion of phosphorus, which could thus contribute to a reduction in environmental pollution.

6 Conclusions

Application of exogenous enzymes as dietary additives for ruminants on fibrous tropical forages and crop residues diets is an evolving technology. This is encouraged by the wide availability of potentially appropriate roughage feedstuffs, the development of better techniques for assessing enzyme activity, and recent advances in biotechnology that have provided more enzymes for cost-effective applications in animal feeding. Exogenous enzymes can play an important role in enhancing animal feed utilization and potentially improve fodder and tree foliage quality in ruminant diets. They can also stimulate the growth of useful bacteria, increase fiber digestion, and improve the capability of rumen bacteria to better degrade diets and secondary metabolites. Enzymes may also reduce fecal nitrogen excretion and creatinine production. Through their effects on diet digestion and rumen functioning, exogenous enzymes added to ruminant diets can lead to improvements in production efficiency and the quality of both meat and milk.

References

Arriola K, Kim S, Staples C, Adesogan A (2011) Effect of fibrolytic enzyme application to low- and high-concentrate diets on the performance of lactating dairy cattle. J Dairy Sci 94:832–841

Awawdeh M, Obeidat B (2011) Effect of supplemental exogenous enzymes on performance of finishing Awassi lambs fed olive cake-containing diets. Livestock Sci 138:20–24

Beauchemin K, Holtshausen L (2010) Developments in enzyme usage in ruminants. In: Enzymes in farm animal nutrition. CABI, pp 206–230

Beauchemin K, Yang W, Rode L (1999) Effects of grain source and enzyme additive on site and extent of nutrient digestion in dairy cows. J Dairy Sci 82:378–390

Beauchemin K, Rode L, Maekawa M et al (2000) Evaluation of a nonstarch polysaccharidase feed enzyme in dairy cow diets. J Dairy Sci 83:543–553

Beauchemin K, Colombatto D, Morgavi D et al (2003) Use of exogenous fibrolytic enzymes to improve feed utilization by ruminants. J Anim Sci 81:E37–E47

Beauchemin K, Kreuzer M, O'Mara F, McAllister T (2008) Nutritional management for enteric methane abatement: a review. Aust J Exp Agric 48:21–27

Bedford M, Cowieson A (2012) Exogenous enzymes and their effects on intestinal microbiology. Anim Feed Sci Technol 173:76–85

Bernard J, Castro J, Mullis N et al (2010) Effect of feeding alfalfa hay or Tifton 85 bermudagrass haylage with or without a cellulase enzyme on performance of Holstein cows. J Dairy Sci 93:5280–5285

Bowman G, Beauchemin K, Shelford J (2002) The proportion of the diet to which fibrolytic enzymes are added affects nutrient digestion by lactating dairy cows. J Dairy Sci 85:3420–3429

Buendía R, González M, Mendoza M et al (2014) Effects of exogenous phytase on growth performance of weaned Dorper x Pelibuey lambs. Anim Nutr Feed Technol 14:183–188

Cheng K-J, McAllister T (1997) Compartmentation in the rumen. In: The rumen microbial ecosystem. Springer, pp 492–522

Chung Y-H, Zhou M, Holtshausen L et al (2012) A fibrolytic enzyme additive for lactating Holstein cow diets: ruminal fermentation, rumen microbial populations, and enteric methane emissions. J Dairy Sci 95:1419–1427

Cruywagen C, Van Zyl W (2008) Effects of a fungal enzyme cocktail treatment of high and low forage diets on lamb growth. Anim Feed Sci Technol 145:151–158

DeFrain J, Hippen A, Kalscheur K et al (2005) Effects of dietary α-amylase on metabolism and performance of transition dairy cows. J Dairy Sci 88:4405–4413

Elghandour MMY, Salem AZM, Gonzalez-Ronquillo M et al (2013) Effects of exogenous enzymes on in vitro gas production kinetics and ruminal fermentation of four fibrous feeds. Anim Feed Sci Technol 179:46–53

Eun J-S, Beauchemin K (2005) Effects of a proteolytic feed enzyme on intake, digestion, ruminal fermentation, and milk production. J Dairy Sci 88:2140–2153

Eun J-S, Beauchemin K, Schulze H (2007) Use of exogenous fibrolytic enzymes to enhance in vitro fermentation of alfalfa hay and corn silage. J Dairy Sci 90:1440–1451

Faniyi TO, Adegbeye MJ, Elghandour MMY et al (2019) Role of diverse fermentative factors towards microbial community shift in ruminants. J Appl Microbiol 30:1–10
Ferraretto L, Shaver R, Espineira M et al (2011) Influence of a reduced-starch diet with or without exogenous amylase on lactation performance by dairy cows. J Dairy Sci 94:1490–1499
Gado H, Salem A, Robinson P, Hassan M (2009) Influence of exogenous enzymes on nutrient digestibility, extent of ruminal fermentation as well as milk production and composition in dairy cows. Anim Feed Sci Technol 154:36–46
Gado H, Salem A, Odongo N et al (2011) Influence of exogenous enzymes ensiled with orange pulp on digestion and growth performance in lambs. Anim Feed Sci Technol 165:131–136
Gado H, Salem A, Camacho L et al (2013) Influence of exogenous enzymes on in vitro ruminal degradation of ensiled rice straw with DDGS. Anim Nutr Feed Technol 13:569–574
Holtshausen L, Chaves A, Beauchemin K et al (2009) Feeding saponin-containing *Yucca schidigera* and *Quillaja saponaria* to decrease enteric methane production in dairy cows. J Dairy Sci 92:2809–2821
Holtshausen L, Chung Y-H, Gerardo-Cuervo H et al (2011) Improved milk production efficiency in early lactation dairy cattle with dietary addition of a developmental fibrolytic enzyme additive. J Dairy Sci 94:899–907
Jalilvand G, Naserian A, Kebreab E et al (2008) Rumen degradation kinetics of alfalfa hay, maize silage and wheat straw treated with fibrolytic enzymes. Arch Zootec 57:155–164
Jiménez-Peralta F, Salem A, Mejia-Hernández P et al (2011) Influence of individual and mixed extracts of two tree species on in vitro gas production kinetics of a high concentrate diet fed to growing lambs. Livestock Sci 136:192–200
Kamble RD, Jadhav AR (2012) Isolation, identification and screening of potential cellulase-free xylanase producing fungi and its production. Afr J Biotechnol 11:14175–14181
Krause DO, Denman SE, Mackie RI et al (2003) Opportunities to improve fiber degradation in the rumen: microbiology, ecology, and genomics. FEMS Microbiol Rev 27:663–693
Lee B, Pometto AL, Demirci A et al (1998) Media evaluation to produce microbial enzymes. J Agric Food Chem 46:4775–4778
López-Aguirre D, Hernández-Meléndez J, Rojo R et al (2016a) Effects of exogenous enzymes and application method on nutrient intake, digestibility, and growth performance of Pelibuey lambs. Springerplus 5:1399
López-Aguirre D, Hernández-Meléndez J, Rojo R et al (2016b) In vitro gas production kinetics and degradability of a diet for growing lambs: effect of fibrolytic enzyme products at different dose levels. J Anim Sci 15:453–460
McAllister T, Oosting S, Popp J et al (1999) Effect of exogenous enzymes on digestibility of barley silage and growth performance of feedlot cattle. Can J Anim Sci 79:353–360
McAllister T, Hristov A, Beauchemin K et al (2001) Enzymes in ruminant diets. In: Enzymes in farm animal nutrition. CABI, pp 273–298
McGinn S, Beauchemin K, Coates T et al (2004) Methane emissions from beef cattle: effects of monensin, sunflower oil, enzymes, yeast, and fumaric acid. J Anim Sci 82:3346–3356
Meale SJ, Beauchemin KA, Hristov A et al (2014) Board-invited review: opportunities and challenges in using exogenous enzymes to improve ruminant production. J Anim Sci 92:427–442
Mendoza GD, Loera-Corral O, Plata-Pérez FX et al (2014) Considerations on the use of exogenous fibrolytic enzymes to improve forage utilization. Sci World J Art 2014:1
Miller D, Elliott R, Norton B (2008a) Effects of an exogenous enzyme, Roxazyme® G2, on intake, digestion and utilisation of sorghum and barley grain-based diets by beef steers. Anim Feed Sci Technol 145:159–181
Miller D, Granzin B, Elliott R et al (2008b) Effects of an exogenous enzyme, Roxazyme® G2 liquid, on milk production in pasture fed dairy cows. Anim Feed Sci Technol 145:194–208
Mora-Jaimes G, Bárcena-Gama R, Mendoza-Martínez G et al (2002) Performance and ruminal fermentation in lambs fed sorghum grain treated with amylases. Agrociencia (Montecillo) 36:31–39

Morgavi D, Beauchemin K, Nsereko V et al (2001) Resistance of feed enzymes to proteolytic inactivation by rumen microorganisms and gastrointestinal proteases. J Anim Sci 79:1621–1630
Morgavi DP, Kelly W, Janssen P et al (2013) Rumen microbial (meta) genomics and its application to ruminant production. Animal 7:184–201
Nozière P, Besle JM, Martin C et al (1996) Effect of barley supplement on microbial fibrolytic enzyme activities and cell wall degradation rate in the rumen. J Sci Food Agric 72:235–242
Peters A, Lebzien P, Meyer U et al (2010) Effect of exogenous fibrolytic enzymes on ruminal fermentation and nutrient digestion in dairy cows. Arch Anim Nutr 64:221–237
Polderman L, Dillon C, Steele A (1955) Why monoethanolamine solution breaks down in gas-treating service. Oil Gas J 54:180–183
Rivero N, Salem A, Gado H et al (2012) Effect of exogenous enzymes and *Salix babylonica* extract or their combination on haematological parameters in growing lambs. J Anim Feed Sci 21:577–586
Rojo R, Kholif A, Salem A et al (2015) Influence of cellulase addition to dairy goat diets on digestion and fermentation, milk production and fatty acid content. J Agric Sci 153:1514–1523
Salem AZM, EL-adawy M, Gado H et al (2011) Effect of exogenous enzymes on nutrient digestibility and growth performance in sheep and goat. Trop Subtrop Agroecosystem 14:867–874
Salem A, Hassan A, Khalil M et al (2012) Effects of sun-drying and exogenous enzymes on nutrients intake, digestibility and nitrogen utilization in sheep fed *Atriplex halimus* foliages. Anim Feed Sci Technol 171:128–135
Salem AZ, Gado H, Colobatto D et al (2013) Effects of exogenous enzymes on nutrient digestibility, ruminal fermentation, and growth performance in beef steers. Livest Sci 154:69–73
Salem AZ, Kholif AE, Olivares M et al (2014) Influence of *S. babylonica* extract on feed intake, growth performance and diet in vitro gas production profile in young lambs. Trop Anim Health Prod 46:213–219
Salem AZM, Alsersy H, Camacho LM et al (2015a) Feed intake, nutrient digestibility, nitrogen utilization, and ruminal fermentation activities in sheep fed *Atriplex halimus* ensiled with three developed enzyme cocktails. Czech J Anim Sci 60:185–194
Salem A, Ammar H, Kholif A et al (2015b) Effect of glucoamylase enzyme extract on in vitro gas production and degradability of two diets with 25% of corn or sorghum grains. Ind J Anim Sci 85:183–188
Salem AZ, Elghandour MM, Kholif AE et al (2015c) The effect of feeding horses a high fiber diet with or without exogenous fibrolytic enzymes supplementation on nutrient digestion, blood chemistry, fecal coliform count, and in vitro fecal fermentation. J Equine Vet Sci 35:735–743
Stella A, Paratte R, Valnegri L et al (2007) Effect of administration of live *Saccharomyces cerevisiae* on milk production, milk composition, blood metabolites, and faecal flora in early lactating dairy goats. Small Ruminant Res 67:7–13
Sujani S, Seresinhe R (2015) Exogenous enzymes in ruminant nutrition: a review. Asian J Anim Sci 9:85–99
Sutton J, Phipps R, Beever D et al (2003) Effect of method of application of a fibrolytic enzyme product on digestive processes and milk production in Holstein-Friesian cows. J Dairy Sci 86:546–556
Titi H (2003) Replacing soybean meal with sunflower meal with or without fibrolytic enzymes in fattening diets of goat kids. Small Ruminant Res 48:45–50
Titi H, Lubbadeh W (2004) Effect of feeding cellulase enzyme on productive responses of pregnant and lactating ewes and goats. Small Ruminant Res 52:137–143
Togtokhbayar N, Cerrillo MA, Rodríguez GB et al (2015) Effect of exogenous xylanase on rumen in vitro gas production and degradability of wheat straw. Anim Sci J 86:765–771
Valdes K, Salem A, Lopez S et al (2015) Influence of exogenous enzymes in presence of *Salix babylonica* extract on digestibility, microbial protein synthesis and performance of lambs fed maize silage. J Agric Sci 153:732–742
Vallejo L, Salem A, Camacho L et al (2016) Effects of xylanase supplementation on feed intake, digestibility, and ruminal fermentation in Rambouillet sheep. J Agric Sci 154:1110–1117

Wang Y, McAllister T, Rode L et al (2001) Effects of an exogenous enzyme preparation on microbial protein synthesis, enzyme activity and attachment to feed in the rumen simulation technique (Rusitec). Brit J Nutr 85:325–332
Yang W, Beauchemin K, Rode L (1999) Effects of an enzyme feed additive on extent of digestion and milk production of lactating dairy cows. J Dairy Sci 82:391–403
Zhang YHP, Lynd LR (2004) Toward an aggregated understanding of enzymatic hydrolysis of cellulose: noncomplexed cellulase systems. Biotechnol Bioeng 88:797–824
Zhou M, Chung YH, Beauchemin K et al (2011) Relationship between rumen methanogens and methane production in dairy cows fed diets supplemented with a feed enzyme additive. J Appl Microbiol 111:1148–1158

Exogenous Fibrolytic Enzymes: For the Better Utilization of *Guinea* Grass and Rice Straw as Ruminant Feeds

Thakshala Seresinhe, Sathya Sujani, and Indunil Pathirana

Abstract Previous in vitro experiments had been conducted to evaluate the effect of exogenous enzymes, cellulase and xylanase, on rumen fermentation parameters of guinea grass ecotype "A" (*Panicum maximum*) and rice straw (*Oryza sativa*). This chapter describes subsequent in vivo studies with female growing goats. The effects of supplementing a Guinea grass–based diet with exogenous cellulase (Experiment 1) and adding exogenous xylanase to a total mixed ration based on Guinea grass and rice straw (Experiment 2) on digestibility, growth performance, and enteric methane production were examined. In both experiments, average daily gain (ADG) in the enzyme-supplemented group was significantly higher than that of the control group. Intake and apparent digestibility of dry matter (DM) and neutral detergent fiber (NDF) were improved significantly. Urinary creatinine and allantoin concentrations significantly increased with cellulase supplementation. The CH_4 production did not differ between enzyme-supplemented diets and the control diets. It can be concluded that supplementation of diets with exogenous cellulase and xylanase for goats was beneficial because of increased ADG, intake, and digestibility of DM and NDF.

1 Introduction

Ruminant livestock play an important role as an integral part of small-scale livestock systems in the world and especially in tropical countries, such as Sri Lanka. Tropical ruminant animal production systems are characterized by low productivity, mainly due to poor-quality feeds, lack of genetic selection, and prevalence of tropical animal diseases (Seresinhe et al. 2012). Most of these production systems, including those in Sri Lanka, depend on permanent year-round grazing on natural pasture or the feeding of agro-industrial by-products (Seresinhe and Pathirana

T. Seresinhe (✉) · S. Sujani · I. Pathirana
Department of Animal Science, Faculty of Agriculture, University of Ruhuna, Kamburupitiya, Sri Lanka

A. Z. M. Salem et al. (eds.), *Exogenous Enzymes as Feed Additives in Ruminants*, https://doi.org/10.1007/978-3-031-27993-5_4

2008). In many areas of Sri Lanka, ruminants depend on year-round grazing on Guinea grass ecotype "A" (*Panicum maximum*). This variety is characterized for having a low digestibility and low feeding value due to high fiber and lignin contents and low crude protein level (Wang et al. 2010). Rice is the most extensively grown cereal crop, and the staple food in Sri Lanka. Its by-product, rice straw, provides an important traditional feed source for ruminants, especially as a potential solution for the dry-season feed shortage (Wang et al. 2010). Feeding rice straw to ruminants is constrained due to several factors such as low digestibility, low protein, low palatability, high-crude fiber (Shen et al. 1998), and high silica content (5–15%). Low digestibility and crude protein levels leading to poor animal performance and productivity are major problems associated with above-mentioned feed sources (Aganga and Tshwenyane 2004).

Enteric methane production is influenced by several factors, including the amount of fermentable carbohydrate and level of fiber in the diet (Eun et al. 2004). Na et al. (2013) demonstrated, in dairy cows, that the higher the fiber content in feedstuffs, the higher was the enteric methane production. Enhancing the digestibility of the fiber component of ruminant feedstuffs has been a huge research interest among ruminant nutritionists, Due to this various chemical, physical, and biological treatment methods have been introduced and some are being practically implemented. Among these approaches, the use of microbial enzymes, such as cellulases and xylanases, has drawn considerable research attention. With the subsequent positive results, as demonstrated by in vitro studies (Sujani et al. 2015) on rumen fermentation parameters, and in vivo studies on rumen fermentation, feed intake, and milk production (Gado et al. 2009), exogenous enzymes are now on the verge of delivering their practical benefits to the ruminant livestock industry. Therefore, following a series of in vitro studies (Sujani et al. 2016, 2017), two in vivo experiments were carried out to investigate the effects of supplementing poor-quality roughages with exogenous fibrolytic enzymes on intake, growth, digestibility, and enteric methane production in goats. The objective of the first experiment was to determine the effect of exogenous cellulase enzyme supplementation on digestibility, growth performance, and enteric methane production in crossbred goats fed with a diet based on guinea grass ecotype "A" (*Panicum maximum*), while the second experiment determined the effects of exogenous xylanase treatment of rice straw incorporated into a total mixed ration (TMR) fed to growing crossbred goats on growth performance, diet digestibility, and enteric methane emission.

2 Material and Methods

2.1 Location and Ethical Approval of Experiments

The experiments were carried out at the Faculty of Agriculture, University of Ruhuna, Kamburupitiya, Sri Lanka. The experimental procedures for the first and second experiments were approved by the Research Ethics Committee of the University of Ruhuna.

2.2 Animals, Management and Enzyme Treatment

2.2.1 Experiment 1

Eight female crossbred (Jamnapari x non-descriptive Sri Lankan) goats of about 18 months old, weighing 19.40 ± 3.03 kg, were used. Goats were first blocked into two groups (four goats per group) on the basis of body weights and thereafter assigned to the treatment (exogenous cellulase enzyme) group (ECE) and to a control (Control) group. Animals were confined to individual metabolism crates (1 m × 0.45 m × 1 m) constructed according to the guidelines of IAEA TECDOC 674 (IAEA 1992) and placed in a well-ventilated shed under uniform management conditions for the whole experimental period. Before the commencement of experiments, all the animals were vaccinated for tetanus (0.5 mL per animal intramuscularly; Serum Institute of India Ltd., Hadaspar, Pune) and dewormed with Ivomectin (0.6 mL per animal subcutaneously; Farm Chemie Pvt. Ltd., Homagama, Sri Lanka).

The goats were fed with fresh guinea grass ecotype "A" (*P. maximum*), harvested daily from a 6-week regrowth on unfertilized (i.e., no addition of inorganic fertilizer) land at the Faculty research farm. The grass was offered ad libitum such that there were at least 10% of daily feed refusals. In addition, the goats received a fixed amount of commercial concentrate (200 g/animal/day, Milk Plus, CIC Holdings, Sri Lanka). The commercial concentrate consisted of cereals and cereal by-products, oilseed meals, vegetable oil, minerals, vitamins, and permitted additives. The chemical composition was as follows in g per 100 g of DM: fat: 6; calcium: 1; phosphorus: 0.75–1.0; metabolizable energy: 10.46 MJ/kg. The goats were also supplemented with common salt (2 g/animal/day, Dumindu Productions, Sri Lanka) and mineral mixture (5 g/animal/day, Farm Chemie Pvt., Sri Lanka). The composition of the mineral mixture was as follows: Ca: 185.2 g; P: 79.0 g; Na: 92.8 g; Mg: 13.6 g; Cu: 1.1 g; Zn: 61.0 g; Mn: 5.3 g; I: 68.0 mg; Co: 21.0 mg; Se: 45.0 mg per kg. Goats were fed with guinea grass supplemented with exogenous cellulose at the dosage of 10 mL/animal/day [this dose is equivalent to 20 μL/g forage dry matter (DM)]. The feeding regime was designed for goats weighing 20–25 kg and expected to attain 25–50 g of average daily gain (ADG), following the recommendations for the daily nutrient requirements of goats (Kearl 1982). The goats were individually fed two times daily at 0930 h and 1530 h and provided with fresh clean water throughout the day.

2.2.2 Experiment 2

Eight female goats of 24 months of age, weighing 23.50 ± 3.77 kg, were used for the second study. Goats were first blocked into two groups (four goats per group) on the basis of body weights and thereafter assigned to the treatment (exogenous xylanase enzyme) group (EXY) and to a control (Control) group. The housing conditions, tetanus vaccination, and worming treatments applied to the goats were the same as those of Experiment 1.

The goats were fed with a TMR formulated with fresh guinea grass ecotype "A" (*Panicum maximum*) (6 weeks regrowth, unfertilized), rice straw (*Oryza sativa*), and commercial concentrate (Milk Plus, CIC Holdings, Sri Lanka) in the following respective proportions (DM basis): 0.52; 0.30; 0.18. The chemical composition of the TMR is presented in Table 4. The TMR was prepared manually each day by mixing chopped guinea grass with rice straw and commercial concentrate. The composition of the commercial concentrate and daily amounts of common salt and mineral mixture given to the goats were the same as those for goats in Experiment 1. To prepare the EXY diet, exogenous xylanase was sprayed and mixed with the rice straw 1 h before the preparation of the TMR. The enzyme dose used was 20 mL/animal/day (40 μL/g of rice straw DM).

As for Experiment 1, the feeding regime was designed for goats weighing 20–25 kg and expected to attain 25–50 g of average daily gain (ADG), following the recommendations for the daily nutrient requirements of goats (Kearl 1982). The goats were individually fed two times daily at 0930 h and 1530 h and provided with fresh clean water throughout the day.

2.3 Experimental Procedure, Measurements and Sampling

The experimental protocols, measurements, and sampling procedures (except for urine sampling) were the same for both experiments. A completely randomized block design was used and within the block, goats were randomly assigned to treatments (ECE in experiment 1 and EXY in experiment 2) or the Control and the same animals were used for both periods within the experiment. The goats were weighed at the beginning of each experiment, and body weight was recorded at 14-day intervals. Each experiment consisted of two 90-day experimental periods, with each period consisting of a 10-day adaptation to the diets, 70-day feeding of experimental diets, and 10-day sampling. During first 7 days of sampling, samples of urine were collected at 8 am, 12 pm, 4 pm, and 8 pm following the procedures described in IAEA-TEC DOC 945 (IAEA, 1997) and acidified with 10% H_2SO_4 to keep the final pH below 3 and then made up to a volume of 100 mL with tap water and stored at −20 °C until analyzed for allantoin and creatinine. Feed offered, feed refusals, and total feces of individual goats were collected and subsamples (about 10% from total measured) were composited across sampling times for each goat. Guinea grass and commercial concentrate were sampled weekly to determine DM content and diets were adjusted weekly to account for changes in DM content.

Methane was measured in the last 3 days of each experimental period. A flexible rubber tube (1 mm wall thickness and 4 mm internal diameter) with 12 holes of 3 mm diameter at the tube head was inserted through the esophagus to the rumen to a depth of approximately 80–100 cm. Rumen gas samples of 100 mL were drawn through the stomach tube and trapped in a specially prepared apparatus (Fig. 1) (100 mL conical flask with an inlet from where a gas sample was inserted and an outlet connected to a gas analyzer; Crowcon Gas Pro, UK). The percentage of CH_4

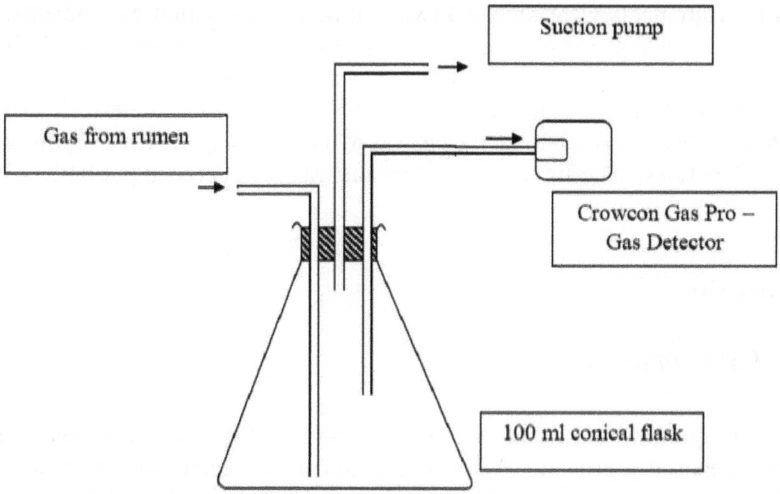

Fig. 1 Apparatus for measuring methane percentage

in the collected gas was recorded on the gas analyzer. Before each measurement, the conical flask was evacuated to ensure that the reading is associated only with the particular gas sample. All the data were recorded at 3 h postfeeding. At least three gas samples from each animal were analyzed at each data recording (Ramos-Morales et al. 2014).

2.4 *Laboratory Analyses*

For both experiments, the DM contents of feed samples, feed refusals, and feces were measured by drying at 55 °C for 48 h. Ground samples (1 mm) were analyzed for ash (525 °C, 3.5 h in muffle furnace, AOAC 1990), crude protein (CP, Kjeltec System 1002, Tecator AB, Hoganas, Sweden) (AOAC 1990), and neutral detergent fiber (NDF) (Van Soest et al. 1991). Preserved urine samples were thawed and diluted with known amounts of distilled water to enable the samples to fall within the concentration range of standards (10–50 mg/L) for the analysis of both allantoin and creatinine. Urinary allantoin and creatinine were determined colorimetrically following the procedures described in IAEA TECDOC 945 (IAEA 1997).

2.5 *Calculations and Statistical Analysis*

Digestibility values were calculated from the daily amounts of DM and nutrients ingested and excreted in feces. Feed conversion ratio was calculated as a ratio of total kilograms of DMI to total kilograms of live weight gain.

As measurements were repeated twice in both trials within experiments, eight replicates per each group were accounted for in the statistical analysis. All the data were subjected to standard analysis of variance (ANOVA) using the general linear model of IBM SPSS Statistics 20.0 (IBM Corporation, Somers, NY, USA). Significance between individual means was identified using the least significant difference (LSD) test. The significance of means was considered at $p < 0.05$.

3 Results

3.1 Experiment 1

Chemical composition of diet, ingredients, the effect of enzyme supplements on intake, digestibility, growth and rumen gas methane concentration, and urinary creatinine and allantoin concentrations are presented in Tables 1, 2, and 3, respectively.

Table 1 Nutrient composition[a] of diets fed to goats (mean ± SEM)

		0.254		
Feeds	DM (%)	OM	CP	NDF
Guinea grass	23.8 ± 1.67	90.6 ± 0.22	9.6 ± 0.14	66.9 ± 1.03
Commercial concentrate	92.9 ± 0.38	88.8 ± 0.17	15.7 ± 0.21	56.2 ± 0.53

[a]*DM* dry matter, *OM* organic matter, *CP* crude protein, *NDF* neutral detergent fiber Values are means of 6 replicates

Table 2 Average daily body weight gain (ADG), intake, and digestibility of nutrients in goats in response to the supplementation of Guinea grass *(Panicum maximum)* with exogenous fibrolytic enzyme, cellulase (ECE) (mean ± SEM of 8 replicates)

Item	ECE[a]	Control	P value
ADG (g/d)	74.0 ± 5.39	60.3 ± 4.38	0.037
Intake (g/d)			
DM	662.8 ± 17.31	611.3 ± 18.38	0.029
OM	602.9 ± 12.68	548.7 ± 13.73	0.016
CP	98.3 ± 1.56	96.5 ± 4.41	0.686
NDF	425.4 ± 11.13	385.5 ± 12.60	0.012
Apparent digestibility (%)			
DM	61.3 ± 2.18	55.6 ± 2.15	0.049
OM	63.0 ± 2.38	57.8 ± 2.20	0.085
CP	78.0 ± 1.60	77.7 ± 1.01	0.887
NDF	64.7 ± 1.97	58.4 ± 2.04	0.016
Feed conversion ratio[b]	9.1	10.6	0.148
CH_4, (mL/100 mL)[c]	10.4 ± 2.04	11.9 ± 1.83	0.528

[a]*ECE* exogenous cellulase, *ADG* average daily gain, *DM* dry matter, *OM* organic matter, *CP* crude protein, *NDF* neutral detergent fiber
[b]Total kilograms of dry matter intake (DMI)/total kilograms of live weight gain
[c]Concentration of methane in samples rumen gas 3 h post-feeding

Table 3 Allantoin and creatinine concentrations in urine from goats fed with Guinea grass (*Panicum maximum*) with supplementary exogenous fibrolytic enzyme cellulase (ECE) and without enzyme supplementation (mean ± SEM of 4 replicates)

Purine derivative (mmol/L)	ECE	Control	P value
Creatinine	4.45 ± 0.27	3.67 ± 0.19	0.025
Allantoin	7.54 ± 0.84	4.95 ± 0.59	0.0003
A: C ratio	1.694 ± 0.13	1.349 ± 0.21	0.027

Table 4 Proximate composition of individual feeds and total mixed ration (TMR) diet (mean ± SEM)

		As a percentage of DM		
Feeds	DM (%)	OM (%)	CP (%)	NDF (%)
Guinea grass	23.8 ± 1.67	90.6 ± 0.17	9.6 ± 0.09	66.9 ± 12
Rice straw	92.4 ± 1.28	87.9 ± 0.86	4.3 ± 0.58	71.0 ± 1.14
Concentrate	92.9 ± 0.44	88.8 ± 0.23	15.7 ± 0.18	56.2 ± 0.49
TMR	33.0 ± 0.51	81.9 ± 0.14	9.1 ± 0.03	60.0 ± 0.76

DM dry matter, *OM* organic matter, *CP* crude protein, *NDF* neutral detergent fiber Values are means of 6 replicates

3.2 *Experiment 2*

Supplementation of rice straw with exogenous xylanase (EXY) significantly increased ADG (g/d) by 42% in goats receiving a TMR compared with the Control subjects receiving unsupplemented rice straw in the TMR. Among intake parameters only the NDF exhibited a statistically significant improvement. Xylanase treatment of rice straw significantly increased the digestibility of DM, OM, and NDF in goats receiving the TMR. Importantly, NDF digestibility showed an 18% increase with EXY supplementation (Tables 4 and 5).

4 Discussion

4.1 *Experiment 1*

Results indicate that supplementation of diets with exogenous cellulase increased ADG (Table 2) nearly by 23% when compared with the control group, probably due to the increases in DMI, DMD, and NDFD. As Lewis et al. (1999) postulated, improved ADG in goats could be also explained by an increase in the supply of available nutrients to animals for deposition and growth. Increased DMD in goats could be also related to the improvement in ruminal fermentation activities, by supplementing guinea grass with exogenous cellulose. In the current study, DMI ranged between 2.51 and 2.56% of body weight (BW), which is less than the reported DMI requirement in tropical goats of 3.0–3.1% of BW (Devendra and McLeroy 1982).

Table 5 Average daily weight gain (ADG), intake, and digestibility of nutrients in response to the exogenous fibrolytic enzyme, xylanase (EXY), supplementation of rice straw based total mixed ration (mean ± SEM of 08 replicates)

Item	EXY	Control	P value
ADG (g/d)	38.5 ± 3.14	27.1 ± 2.08	0.003
Intake (g/d)			
DM	674.5 ± 26.44	629.1 ± 16.85	0.95
OM	567.8 ± 19.14	531.5 ± 4.29	0.136
CP	62.4 ± 1.80	58.8 ± 1.57	0.089
NDF	406.2 ± 11.14	372.1 ± 12.63	0.003
Apparent digestibility (%)			
DM	64.5 ± 0.68	60.4 ± 1.17	0.001
OM	69.7 ± 0.58	66.6 ± 0.61	0.001
CP	71.7 ± 0.87	69.7 ± 1.02	0.079
NDF	57.1 ± 0.83	49 ± 0.84	0.003
CH_4 (mL/100 mL)[a]	11.8 ± 0.67	12.4 ± 1.68	0.711

EXY exogenous xylanase, *ADG* average daily gain, *DM* dry matter, *OM* organic matter, *CP* crude protein, *NDF* neutral detergent fiber

[a]Concentration of methane in rumen gas samples, 3 h postfeeding

As stated by Kearl (1982), DMI of a goat (average weight 25 kg) for growth (for a 50–75 g/d ADG) and maintenance should be 710–730 g/d. Despite having a slightly lower DMI, both the ECE and Control groups attained the ADG target of 50 g/d; this might have resulted from the significantly enhanced DMD and NDFD values. In agreement with the present results, exogenous enzymes have been reported to increase ADG, DMI, and DMD in several studies where small ruminants have been used (Salem et al. 2011). On the contrary, Awawdeh and Obeidat (2011) reported that fibrolytic enzymes had no significant influence on feed intake, ADG, and nutrient digestibility in goats and lambs. Although cellulase supplementation of guinea grass and cereal concentrate fed to goats in the present study resulted in significantly higher ADG and DMD values, feed conversion ratio was not affected. However, Beauchemin et al. (1997) reported 10% and 11% improvements, respectively, in the feed conversion ratio of steers fed dry forages and feedlot finishing cattle on concentrate diets.

Since the CP content of guinea grass is low, a significant change in CP intake and digestibility was expected. However, the lack of an effect of enzyme supplementation on CP intake and CP digestibility could most probably be due to the inclusion of a highly digestible commercial concentrate in the diet, which supplies most of the dietary protein. Yang et al. (2000) concluded that exogenous fibrolytic enzymes appeared to improve digestion only when fiber digestion is restricted, but no effects occur when digestibility is high, like in commercial concentrate. In the present study, CP intake in goats was 96.55–98.32 g/d in control and ECE groups, respectively. The digestibility of CP in the present study (81.19–81.62% in control and ECE groups, respectively) is higher than the values reported by Titi and Lubbadeh (2004) where they observed 72.1–73.5% of CP digestibility in Awasssi lambs fed with a cellulase-supplemented diet. In contrast to the higher CP digestibility values reported by Titi and Lubbadeh (2004) studies, Gonzalez et al. (2008) obtained comparatively lower values (59.6–63%) in dairy goats supplemented with fibrolytic enzymes.

As suggested by Dado and Allen (1995), increased NDF digestibility may be the main cause for the significantly enhanced NDF intake and DMI, by reducing the physical rumen fill. In the present study, NDF digestibility increased approximately by 9% and the result agrees with the findings of Salem et al. (2011) and Ganai et al. (2011). Oba and Allen (1999) found that a % increase in NDF degradation, measured in vitro, could be associated with a 0.17 kg increase in DMI and 0.25 kg increase in 4% fat-corrected milk yield. Thus, the present results suggest that cellulase supplementation would be very much beneficial in enhancing DMI and milk production in goats.

In the present study, urinary allantoin was estimated as it has been identified as the major purine component. Carro et al. (2012) reported that allantoin comprised 87.3–90.8% of total purine derivative excretion in goats, in agreement with earlier results for this species (Jetana, 2005). Purwati et al. (2013) reported that the proportions of total urinary purine derivatives represented by xanthine and hypoxanthine in goats were very low (0.32 and 0.36%, respectively). Therefore, the excretion of xanthine and hypoxanthine in the present experiment was not estimated. Peer-reviewed published data on urinary allantoin excretion in sheep or in goats supplemented with fibrolytic enzymes are extremely limited. Elliot and Topps (1963) found that the lower the protein content of the diet, the less the output of urea, ammonia, uric acid, hippuric acid, and, in general, allantoin. It can therefore be assumed that the significantly enhanced excretion of allantoin was a result of high dietary protein content, even though in the present study, CP intake did not significantly vary between treatment and control groups. The response in allantoin concentration to cellulase supplementation may be attributed to the increased DMI and presumably to the improved apparent digestibility. It is generally presumed that the excretion rate of creatinine is relatively constant in healthy animals and remains independent of level of feed intake. Brody (1964) stated that creatinine is an indicator of body protein turnover and creatinine in urine is excreted in proportion to live weight. Since significant increases in ADG and final body weight were observed in the goats receiving exogenous cellulase supplementation, the concurrent increases in urinary creatinine concentrations were to be expected. Lamothe et al. (2002) indicated that urinary allantoin has a potential as a tool to predict bacterial protein production in ruminants. They confirmed that bacterial CP predicted by using allantoin as a marker was well related to bacterial crude protein (BCP) predicted by the NRC model. Further, they observed that allantoin: creatinine ratio was increased, when the diet digestibility increased during the wet season and vice versa. Therefore, it can be suggested that in the present study, higher apparent digestibility ($P < 0.5$) coupled with higher allantoin: creatinine ratio in goats fed with enzyme-treated diet could have been associated with higher BCP production. George et al. (2011) reported that values for urinary allantoin concentrations ranged from 4.68 to 7.40 mmol/L in a study where Barberi goats were fed with a 50: 50 forage/concentrate diet; these results were consistent with the urinary allantoin concentrations observed in the present study. In the study of George et al. (2011), creatinine concentrations ranged from 4.02 to 4.20 mmol/L, which were slightly higher than the values observed in the urine from the control goats in our study. In contrast to these values, Marapana and Seresinhe (2007) obtained much lower values ranging from 1.001 to 2.910 and 1.110 to 2.590 mmol/L for allantoin and creatinine in goats fed with a farm and an experimental diet, respectively.

The number of studies, which have reported the effect of fibrolytic enzyme supplementation on CH_4 production, is limited. Most of them have focused on in vitro trials and very few are on in vivo experiments. The lack of effect of cellulase supplementation on rumen gas CH_4 concentrations observed in the present study concurs with the findings of Giraldo et al. (2007) who found that methane production was not influenced by cellulase supplementation in a diet composed of 70% forage and 30% concentrate (DM basis). Soltan et al. (2013) indicated that the use of exogenous enzymes, particularly cellulases, increases CH_4 production, indicating that there is a forage–enzyme interaction. These results are consistent with the findings of Gemeda and Hassen (2015) who observed increasing volume of methane gas with increasing levels of enzyme application for the feeds treated with enzymes, compared to the control samples. Chung et al. (2012) reported that methane production, whether calculated as per kg of DM or per kg of milk, would increase when enzyme was added in dairy diets. Dong et al. (1999) also found methane production increased by 43% when cellulase and xylanase were added in hay. However, in a study conducted by Tang et al. (2013), CH_4 production was decreased with cellulase supplementation. In future research endeavors, attention should be paid to analyze the volatile fatty acid profiles (VFA) and CH_4 production concurrently as the VFA profile is directly related to CH_4 production. A VFA profile shifted toward more acetate production will indicate the increased CH_4 production and this phenomenon was observed by Gado et al. (2009). However, Arriola et al. (2011) observed that in cows fed with high concentrate diets, enzyme supplementation decreased the acetate:propionate ratio in rumen fluid, and hence a reduction in methane production (calculated from VFA profiles, using a fermentation balance approach). The shift in the pattern of VFA is reported to be influenced by the type of diet and enzyme preparations (Giraldo et al. 2008).

4.2 *Experiment 2*

According to the existing body of knowledge, increasing the level of rice straw in the diet leads to an increased dietary NDF content. Several studies have shown that a sharp increase in dietary fiber content invariably led to reduced nutrient digestibility (Dado and Allen 1995; Zhao et al. 2007). Despite the relatively high NDF content in the diet in the present study, improvements in performance and other parameters can be attributed to the positive effect of exogenous xylanase enzymes.

Significantly enhanced ADG can be attributed to the increased NDF intake and digestibility of DM and NDF. The higher NDF digestibility may be the main cause for the enhanced NDF intake by reducing the physical rumen fill (Dado and Allen 1995). A number of studies have examined the effect of fibrolytic enzyme supplementation on NDF digestibility in sheep (van de Vyver and Useni 2012) and goats (Bala et al. 2009). Phakachoed et al. (2012) reported enhancement of NDF digestibility of cereal straw-based diets supplemented with fibrolytic enzymes. Devendra and McLeroy (1982) indicated that optimum DMI for a tropical goat should be just 3.0–3.1% of body weight, but in the current study, the DMI was 2.63–2.81% of body weight. Although the TMR was designed to fulfill the nutrient requirement of goats to achieve a growth rate of 25–50 g/d, none of the groups achieved the upper limit of ADG. The lower ADG was likely due to the DMI being below the upper limit of 720 g/d to achieve a daily gain of 50 g (Kearl 1982). Bhasker et al. (2013) observed the significantly enhanced DM and NDF digestibility in fibrolytic-enzyme-supplemented corn stover-based TMR and sorghum stover-based TMR, respectively, which can be considered similar to the present study, since in all these studies, low-quality roughages have been used. In contrast to present study, Pinos Rodríguez et al. (2002) reported no effect on fiber digestibility in ruminants fed hay or silage-based TMR supplemented with fibrolytic enzymes.

In the present study, the main source of CP was the commercial concentrate with a high digestibility; hence, it can be assumed that CP digestibility and intake was not influenced by the addition of fibrolytic enzymes. As concluded by Yang et al. (2000), exogenous fibrolytic enzymes appear to improve digestion only when fiber digestion is restricted, but no effects occur when digestibility is high, as is the case with a commercial cereal-based concentrate. In the present study, intake of CP is 62.4–58.9 g/animal/day, which is well above the standard dietary CP requirement of 43–48 g/day (Kearl 1982). Salem et al. (2011) observed unchanged CP intake but significantly enhanced CP digestibility in goats fed with high-quality hay supplemented with fibrolytic enzyme. The digestibility of CP in the present study, which was approximately 71% for both groups, is similar to the results obtained by Titi and Lubbadeh (2004) where they observed 72.1–73.5% CP digestibility in Awasssi lambs fed with a fibrolytic-enzyme-supplemented diet. In contrast to the higher CP digestibility values reported by above studies, Gonzalez et al. (2008) obtained comparatively lower values (59.6–63%) in dairy goats supplemented with fibrolytic enzymes.

The number of studies where the effect of supplementing diets with exogenous fibrolytic enzymes on rumen methane production has been considered is limited, with most being conducted under in vitro conditions. The results of the present study concur with those of Giraldo et al. (2007) who found that methane production was not influenced by fibrolytic enzyme supplementation. In an in vitro study conducted by Togtokhbayar et al. (2015) where they have used wheat straw supplemented with Xylanase (same product which was used in this experiment) in goat rumen fluid, they observed increased CH_4 production, linearly at 24 h of incubation, and then unaffected at 6 and 48 h of incubation.

5 Conclusions

- Supplementation of a guinea grass–based diet with exogenous fibrolytic cellulase enzyme fed to female goats resulted in increases in ADG, intake and digestibility of DM and NDF, and urinary allantoin concentration. However, enteric methane production was not significantly affected.
- Supplementation of a Guinea grass/rice straw-based diet with exogenous xylanase enzyme enhanced the ADG, NDF intake, and digestibilities of DM, OM, and NDF in female goats.
- It is suggested that further studies should examine the effects of fibrolytic enzyme supplementation on production performance, especially on milk production in ruminants. Moreover, there is a need to investigate the effects of fibrolytic enzyme supplementation on in vivo methane production, using accurate techniques, such as the respiration chamber method.

Acknowledgments Authors are very grateful to the International Atomic Energy Agency for the continuous financial support through IAEA/CRP project on "The use of enzymes and nuclear technologies to improve the utilization of fibrous feeds and reduce greenhouse gas emission from ruminant livestock."

References

Aganga AA, Tshwenyane S (2004) Potentials of guinea grass (*Panicum maximum*) as forage crop in livestock production. Pak J Nutri 3:1–4

AOAC (1990) Official methods of analysis, 15th edn. Association of Official Analytical Chemists, Arlington

Arriola KG, Kim SC, Staples CR, Adesogan AT (2011) Effect of fibrolytic enzyme application to low- and high- concentrate diets on the performance of lactating dairy cattle. J Dairy Sci 94:832–841

Awawdeh MS, Obeidat BS (2011) Effect of supplemental exogenous enzymes on performance of finishing Awassi lambs fed olive cake-containing diets. Livestock Sci 138:20–24

Bala P, Malik R, Srinivas B (2009) Effect of fortifying concentrate supplement with fibrolytic enzymes on nutrient utilization, milk yield and composition in lactating goats. J Anim Sci 80:265–272

Beauchemin KA, Jones SDM, Rode LM et al (1997) Effects of fibrolytic enzyme in corn or barley diets on performance and carcass characteristics of feedlot cattle. Can J Anim Sci 77:645–653

Bhasker TV, Nagalakshmi D, Rao DS (2013) Development of appropriate fibrolytic enzyme combination for maize Stover and its effect on rumen fermentation in sheep. Asian Aust J Anim Sci 26:945–951

Brody S (1964) Bioenergetics and growth. Hafner Publishing Company, New York

Carro MD, Cantalapiedra- Hijar G, Ranilla MJ et al (2012) Urinary excretion of purine derivatives, microbial protein synthesis, N utilization, and ruminal fermentation in sheep and goats fed diets of different quality. J Anim Sci 90:3963–3972

Chung YH, Zhou M, Holtshausen L et al (2012) A fibrolytic enzyme additive for lactating Holstein cow diets: ruminal fermentation, rumen microbial populations and enteric methane emissions. J Dairy Sci 95:1419–1427

Dado RG, Allen MS (1995) Intake limitations, feeding behavior, and rumen function of cows challenged with rumen fill from dietary fibre or inert bulk. J Dairy Sci 78:118–133

Devendra C, McLeroy GB (1982) Goat and sheep production in the tropics. Longman, London

Dong Y, Bae HD, McAllister TA, Mathison GW et al (1999) Effects of exogenous fibrolytic enzymes α-bromoethanesulfonate and monensin on fermentation in a rumen simulation (RUSITEC) system. Can J Anim Sci 79:491–498

Elliott RC, Topps JH (1963) Nitrogen metabolism of African cattle fed diets with an adequate energy, low protein content. Nature 197:668–670

Eun J, Fellner V, Gumpertz ML (2004) Methane production by mixed ruminal cultures incubated in dual-flow fermentors. J Dairy Sci 87:112–121

Gado HM, Salem AZM, Robinson PH et al (2009) Influence of exogenous enzymes on nutrient digestibility, extent of ruminal fermentation as well as milk production and composition in dairy cows. Anim Feed Sci Technol 154(1):36–46

Ganai AM, Sharma T, Dhrla RK (2011) Influence of exogenous fibrolytic enzymes on in vitro fermentation of bajra straw in goats. Vet Pract 12:138–141

Gemeda BS, Hassen A (2015) Methane production of two roughage and total mixed ration as influenced by cellulase and xylanase enzyme addition. Sci Agric 72:11–19

George SK, Verma AK, Mehra UR et al (2011) Evaluation of purine metabolites – creatinine index to predict the rumen microbial protein synthesis from urinary spot samples in Barbari goats. J Anim Feed Sci 20:509–525

Giraldo LA, Tejido ML, Ranilla MJ (2007) Effects of exogenous cellulase supplementation on microbial growth and ruminal fermentation of a high-forage diet in Rusitec fermenters. J Anim Sci 85:1962–1970

Giraldo LA, Tejido ML, Ranilla MJ, Ramos S, Carro MD (2008) Influence of direct-fed fibrolytic enzymes on diet digestibility and ruminal activity in sheep fed a grass hay-based diet. J Anim Sci 86:1617–1623

González E, Cajal G, Albanell E et al (2008) Lactational effects of adding a fibrolytic enzyme complex to the concentrate of lactating dairy goats. J Anim Feed Sci 17:344–351

International Atomic Energy Agency (IAEA) (1992) Manual on measurement of methane and nitrous oxide emissions from agriculture. IAEA-TECDOC-674. Vienna, IAEA

International Atomic Energy Agency (IAEA) (1997) Estimation of rumen microbial protein production from purine derivatives in urine. IAEA-TECDOC-945. Vienna, IAEA

Jetana T (2005) Urinary purine derivatives as index for estimation of ruminal microbial nitrogen production in sheep and goats. Doctoral dissertation. School of Graduate Studies, University Putra, Malaysia

Kearl LC (1982) Nutrient requirements of ruminant in developing countries. International Feedstuff Institute. Utah Agricultural Experiment Station, Logan

Lamothe M, Klopfenstein T, Adams D, Musgrave J, Erickson G (2002) Nebraska Beef Cattle Report. Copyright © 2001 The Board of Regents of the University of Nebraska, Lincoln

Lewis GE, Sanchez WK, Hunt CW et al (1999) Effect of direct-fed fibrolytic enzymes on the lactational performance of dairy cows. J Dairy Sci 82:611–617

Marapana RAUJ, Seresinhe T (2007) Effect of feeding regime on growth, digestibility and excretion of purine derivatives in goats. In: Proceedings of 4th Academic Sessions. University of Ruhuna, Sri Lanka

Na R, Dong H, Zhu Z et al (2013) Effects of forage type and dietary concentrate to forage ratio on methane emissions and rumen fermentation characteristics of dairy cows in China. Agric Biosyst Eng 56:1115–1122

Oba M, Allen MS (1999) Evaluation of the importance of the digestibility of neutral detergent fibre from forage: effects on dry matter intake and milk yield of dairy cows. J Dairy Sci 82:589–596

Phakachoed N, Lounglawan P, Suksombat W (2012) Effects of xylanase supplementation on ruminal digestibility in fistulated non-lactating dairy cows fed rice straw. Livest Sci 149:104–108

Pinos-Rodríguez JM, González SS, Mendoza GD et al (2002) Effect of exogenous fibrolytic enzyme on ruminal fermentation and digestibility of alfalfa and rye-grass hay fed to lambs. J Anim Sci 80:3016–3020

Purwati CS, Yusiati LM, Budhi SPS (2013) The contribution of basal purine excretion on the total excretion of purine derivatives in the urine of Bligon and Kejobong goats. Bull Anim Sci 37:6–11

Ramos-Moralesa E, Arco-Péreza A, Martín-Garcíaa AI et al (2014) Use of stomach tubing as an alternative to rumen cannulation to study ruminal fermentation and microbiota in sheep and goats. Anim Feed Sci Technol 198:57–66

Salem AZM, El-Adawy M, Gado H et al (2011) Effects of exogenous enzymes on nutrients digestibility and growth performance in sheep and goats. Tropic Subtropic Agroecosyst 14:867–874

Seresinhe T, Pathirana KK (2008) Effect of supplementation of straw based diets on the digestibility and microbial nitrogen production of cross bred cattle. World J Agric Sci 4:745–751

Seresinhe T, Madushika SAC, Seresinhe Y et al (2012) Effects of tropical high tannin non-legume and low-tannin legume browse mixtures on fermentation parameters and methanogenesis using gas production technique. Asian Aust J Anim Sci 25:1404–1410

Shen HS, Ni DB, Sundstøl F (1998) Studies on untreated and urea-treated rice straw from three cultivation seasons: 1. Physical and chemical measurements in straw and straw fractions. Anim Feed Sci Technol 73:243–261

Soltan YA, Abdalla AL, Silva LRF et al (2013) Response of different tropical pasture grass species to treatments with fibrolytic enzymes in terms of in vitro ruminal nutrient degradation and methanogenesis. Anim Nutri Feed Technol 13:551–568

Sujani S, Pathirana IN, Seresinhe RT (2015) Enhanced in vitro fermentation parameters of Guinea grass ecotype 'A' (*Panicum maximum*) and rice straw (*Oryza sativa*) with supplementation of exogenous fibrolytic enzymes. Livest Res Rural Dev 27:41

Sujani S, Pathirana IN, Seresinhe RT et al (2016) In vitro effects of exogenous fibrolytic enzymes on rumen fermentation of wild Guinea grass (*Panicum maximum*). Iran J Appl Anim Sci 6:303–308

Sujani S, Piyasena TD, Pathirana IN et al (2017) Supplementation of rice straw (*Oryza sativa*) with exogenous fibrolytic enzymes improves rumen fermentation on in vitro system. Turkish J Anim Vet Sci 41:25–29

Tang SX, Zou Y, Wang M et al (2013) Effects of exogenous cellulase source on in vitro fermentation characteristics and methane production of crop straws and grasses. Anim Nutr Feed Technol 13:489–505

Titi HH, Lubbadeh WF (2004) Effect of feeding cellulase enzyme on productive responses of pregnant and lactating ewes and goats. Small Ruminant Res 52:137–143

Togtokhbayar N, María AC, Germán BR et al (2015) Effect of exogenous xylanase on rumen in vitro gas production and degradability of wheat straw. Anim Sci J 86:765–771

van de Vyver WFJ, Useni BA (2012) Digestion and microbial protein synthesis in sheep as affected by exogenous fibrolytic enzymes. S Afr J Anim Sci 42:488–492

Van Soest PJ, Robertson JB, Lewis BA (1991) Symposium: carbohydrate methodology, metabolism, and nutritional implications in dairy cattle. J Dairy Sci 74:3583–3597

Wang M, Zhaoa XG, Tan SL et al (2010) Effects of increasing level of dietary rice straw on chewing activity, ruminal fermentation and fibrolytic enzyme activity in growing goats. Asian Aust J Anim Sci 23:1022–1027

Yang WZ, Beauchemin KA, Rode LM (2000) A comparison of methods of adding fibrolytic enzymes to lactating cow diets. J Dairy Sci 83:2512–2520

Zhao XG, Jiang HL, Sun ZH et al (2007) Effect of rice straw in the diet for growing goats on site and extent of digestion and N balance. J Anim Feed Sci 16:379–388

Role of Exogenous Enzymes in Feed Digestibility and Reducing the Emission Intensity of Enteric Methane Production in Ruminants

Abubeker Hassen, B. S. Gemeda, K. Selzer, T. Nel, Abdelfattah Zeidan Mohamed Salem, Mona Mohamed Mohamed Yasseen Elghandour, O. M. M. Ahmed, and A. M. Akanmu

Abstract Several ruminant production systems depend on forage as a major source of nutritional constituents. The presence of diversified microflora in the rumen helps ruminants to adapt to a variety of ecological niches, as the microbes were synthesizing and secreting fibrolytic enzyme complex in order to hydrolyze plant fiber. Exogenous cellulases, hemicellulases, xylanase, and pectinases are used in ruminant feed biotechnology to enhance feed utilization, improve digestibility of certain feed components, and increase milk or meat production. At the initial stage of evlauation of these enzymes, in vitro gas production is used as an indication of feed fermentation. Indeed, the degradation of the substrate in the rumen is directly and positively correlated with gas production, and thus in vitro gas production techniques was utilised widely to determine the effect of supplemenation of exogenious enzyme on feed degradation. Nevertheless, in vitro gas production can also be used to estimate the methane emission potential in forges and feedstuffs. The mixture of exogenous enzymes affects the rumen fermentation characteristics, increase volatile fatty acid production and improves the nutrients supply, which can be used to support animal production and eventually reduce the enteric methane per unit of products in ruminants. This chapter highlights the effect of exogenous enzymes on the digestibility of high forage total mixed ration, research techniques used to quanity their effect and also their ability to indirectly mitigate in vitro methane production in livestock by reducing the intensity of methane emission per unit product.

A. Hassen (✉) · B. S. Gemeda · K. Selzer · T. Nel · O. M. M. Ahmed · A. M. Akanmu
Department of Animal Science, Faculty of Natural and Agricultural Sciences, University of Pretoria, Pretoria, South Africa
e-mail: Abubeker.hassen@up.ac.za

A. Z. M. Salem · M. M. M. Y. Elghandour
Facultad de Medicina Veterinaria y Zootecnia, Universidad Autónoma del Estado de México, Toluca, Edo de México, Mexico

A. Z. M. Salem et al. (eds.), *Exogenous Enzymes as Feed Additives in Ruminants*, https://doi.org/10.1007/978-3-031-27993-5_5

1 Introduction

Forages are the bulk component of ruminants' feed, which also serves as the main nutritional component in different ruminant production systems across the world (Wilkins 2000). Ruminants have adapted to a variety of ecological niches because of the huge presence of diverse microbial population that resides in the rumen. Slow to fast forage degradation is made possible by rumen microbes that synthesize and secrete a fibrolytic enzyme complex, capable of hydrolyzing plant fiber (Varga and Kolver 1997). The nutrition of ruminants under extensive system is determined, among other things, by the extent of microbial fermentation that occurs in the foregut or reticulorumen and their adaptation to digestion of high-fiber feeds (Wallace et al. 2001). The rumen, also referred to as fermentation vat, is home to three major groups of microorganisms: bacteria, protozoa, and fungi. A synergistic relationship exists between the host animal and these rumen microorganisms, in which the host provides a constant supply of feed by ingestion and provision of a suitable rumen microenvironment for microbial survival and proliferation. The microorganisms, on the other hand, attack, degrade, and ferment the feed, thereby producing volatile fatty acids as an energy source and microbial cells as a protein source for the host animal (France and Dijkstra 2005). This fermentation enables ruminants to make maximal use of low-quality forages that otherwise cannot be utilized by monogastric species (Wallace et al. 2001). This allows ruminants to convert low-quality feeds into meat, milk, wool, etc., which are the basis of the animal agricultural industry that has served humanity through millennia (Weimer et al. 2009).

Although ruminants can utilize feed of very low quality to survive, production efficiency is not at its best on these low-quality feeds. This is because only a small percentage (10–35%) of the gross energy intake of these low-quality feeds is retained as net energy in the end, because the animal may not be able to digest 20–70% of the cellulose (Marquardt and Brufau 1997). Factors like feed type and animal profile determine the rate and extent of fiber degradation in ruminants (Varga and Kolver 1997). Feed factor includes intake (Tyrrell and Moe 1975), physical and chemical nature of the forage (Jung and Allen 1995), particle size (Yang and Beauchemin 2001), chemical and biological treatment (Zhong et al. 2011), and feeding strategies, and additives (Wenk 2000). A ruminant's intake and digestibility are affected mainly by the passage rate and mean retention time of the feed in its rumen. These factors, in turn, are affected by dietary, animal, and climatic factors.

Dietary factors such as the particle size of feed are known to affect digestibility. The more finely the feed, the faster is the digestion. This is because of the increased surface area of the feed particles for better microbial attachment and colonization, which increases passage rate and ultimately feed intake. Fiber content, though presumed to decrease intake, it increases passage rate when present in relatively small amounts. By stimulating propulsive activity, low fiber content reduces feed retention time in the digestive tract, thus decreasing digestion. But when fiber is present in excessive amounts, it has the reverse effect (Meissner 1997).

Climatic factors such as temperature can also have an effect. A positive relationship exists between temperature and the mean retention time. An increase in mean retention time is expected with an increase in temperature, which leads to a decreased passage rate and decreased intake (Faichney 1986).

Ruminants consuming feed with a higher proportion of concentrate such as grains or starch generally emitted lower CH_4. However, because of the cost implications and concentrates shortages in developing countries, multiple livestock production systems cannot use concentrates as a sustainable methane mitigation strategy. Many researchers have concluded that increasing the quality of forage, rate of fermentation and passage could generally result in the reduction of enteric CH_4 emissions (Boland et al. 2009; McGeough et al. 2010). However, improving the fermentation of poor-quality forage using fibrolytic exogenous enzymes often results in a higher ruminal fermentation characterized by higher acetate production and total volatile fatty acid produced. Higher acetate production means higher hydrogen avilability for methane synthesis though the improvement in fermentation and the digestibility of fiber by ruminants under extensive livestock production may indirectly reduce the intensity of methane emission due to efficient utilization of the feed and improvement in productivity of herds in these production systems that rely on roughage as the bulk of the feed source for ruminants.

2 Poor-quality Roughage and its Degradation by Ruminants

In many countries, the main feedstuffs for ruminant animal production are pastures and crop residue, generally regarded as roughages. Roughages are bulky materials, which have high fiber content (> 18% crude fibre) and low nutrient density. Hay, pasture, silage, straw, and cottonseed hulls are examples of roughages (Birhan and Adugna 2014). There is no doubt that during the dry season, the main basal feeds for ruminants in warm-climates in developing countries are essentially crop residues and poor-quality grasses from rangelands, which are grazed or collected manually at a very advanced vegetation stage.

The fermentation process of crop residues or forage in the rumen begins with the attachment of the microbes to the feed through adhesion. This is followed by successive microbial colonization, which occurs quite rapidly. Thus, the attachment of ruminal microorganisms to their substrate is a prerequisite to the digestion of forage particles in the rumen (Varga and Kolver 1997).

Tropical pastures, which are often of variable quality, are mostly of poor nutritional value at maturity owing to a high rate of lignification (Meissner 1997). Lignification increases with maturity, leading to poorer degradation and lower feed utilization efficiency (Krueger et al. 2008a, b). This inefficiency can result in increasing the quantity of feed that is necessary to satisfy the nutrient need of animals in order to maintain the required levels of animal performance. That then increases feeding costs and environmental pollution because of increased waste per unit of animal product (Sheppy 2001). In periods of high feed costs, increasing the efficiency of feed utilization would decrease the cost of feeding and reduce the pollution of the environment associated with the excretion of unused nutrients such as urea, and loss of energy as methane (CH_4) gas.

In Africa, the main feed resource for livestock in traditional crop-livestock production systems is crop residue (Birhan and Adugna 2014). Ruminants have evolved in a way that enables them to utilize cellulolytic material. However, agricultural systems based on maximum exploitation of these poor quality feed resources result in poor animal production, primarily because of low intake (Ndlovu and Buchanan-Smith 1985). These feeds when used as sole feed will result in malnutrition, which leads to a decrease in productivity and a reduction in disease resistance and overall health of the animal (Birhan and Adugna 2014). These feedstuffs are often unable to support rumen conditions that are conducive to optimal microbial activity because of their deficiency in total nitrogen, true protein, readily fermentable carbohydrates, and minerals. There are many ways by which farmers may improve the value of their poor-quality forages and crop residues. Poor-quality constraints can be overcome to some extent by certain physical, chemical and biological treatments. One such biological treatment involves exogenous enzyme pretreatment.

Too many low-quality roughages are characterized by low-protein and high-fiber content. These poor-quality roughages can be grouped as follows: (i) fodder conserved on the farm such as cereal straw, poor meadow hay, and poor lucerne hay; (ii) native grasses including the edible shrubs or those conserved as hay; (iii) mature natural herbage and stems; and (iv) by-products from farms and factories such as corncobs, corn stalks, cottonseed hulls, old newspapers and sawdust. All these roughage feeds vary in physical make-up and appearance, and are all good sources of energy but are generally lignified and thus less palatable to animals. Animals fed on these low-quality forage will suffer condition and performance loss. Mature stock also will have difficulty in maintaining condition on such feed and it is unsuitable for young growing, pregnant, lactating, and wool-producing stock (Khan et al. 2012).

Six major factors that affect the quality of roughages are as follows:

(i) Maturity: This is the most important factor and some of the other factors that affect forage quality exert indirectly their influence by accelerating or delaying the maturity of plants. As a plant cell ages, the cell wall content increases and causes indigestible lignin to accumulate.
(ii) Crop species: Grasses and legumes vary widely in protein content, with legumes usually being higher.
(iii) Harvest and storage: Improper harvest techniques can lead to the loss of leaves, with higher nutritive value than stems, whereas incorrectly storing hay (e.g., incorrect moisture content) or improperly ensiling forage crops can lower the quality significantly.
(iv) Environment: Environmental factors mainly moisture, temperature, and the amount of sunlight influence forage quality.

(v) Soil fertility: Soil fertility usually affects forage yield much more than it affects the quality. Low and very high soil fertility can reduce forage quality.
(vi) Variety: The variety or type of cultivar can affect the forage quality, but its effect is not so great as the other five factors (Van Soest et al. 1978).

The characteristics of plant material also determine its nutritional value (Walters 1971). This can be described in terms of the structure and composition of primary and secondary plant cell walls (Fig. 1). Roughage consists of a heterogeneous population of cell types, each possessing its own degradability characteristics, which are determined by (i) their location within the plant, (ii) their anatomical features, and (iii) the chemical composition of their walls (Travis et al. 1997). Weimer (1996) stated that the architecture of the plant cell may be just as important as its chemistry. According to Loix et al. (2017) the main building blocks of the primary cell wall are proteins and polysaccharides such as cellulose, hemicellulose, and pectin (Figure 1A). Formation of a secondary cell wall is accomplished by adding phenolic components such as

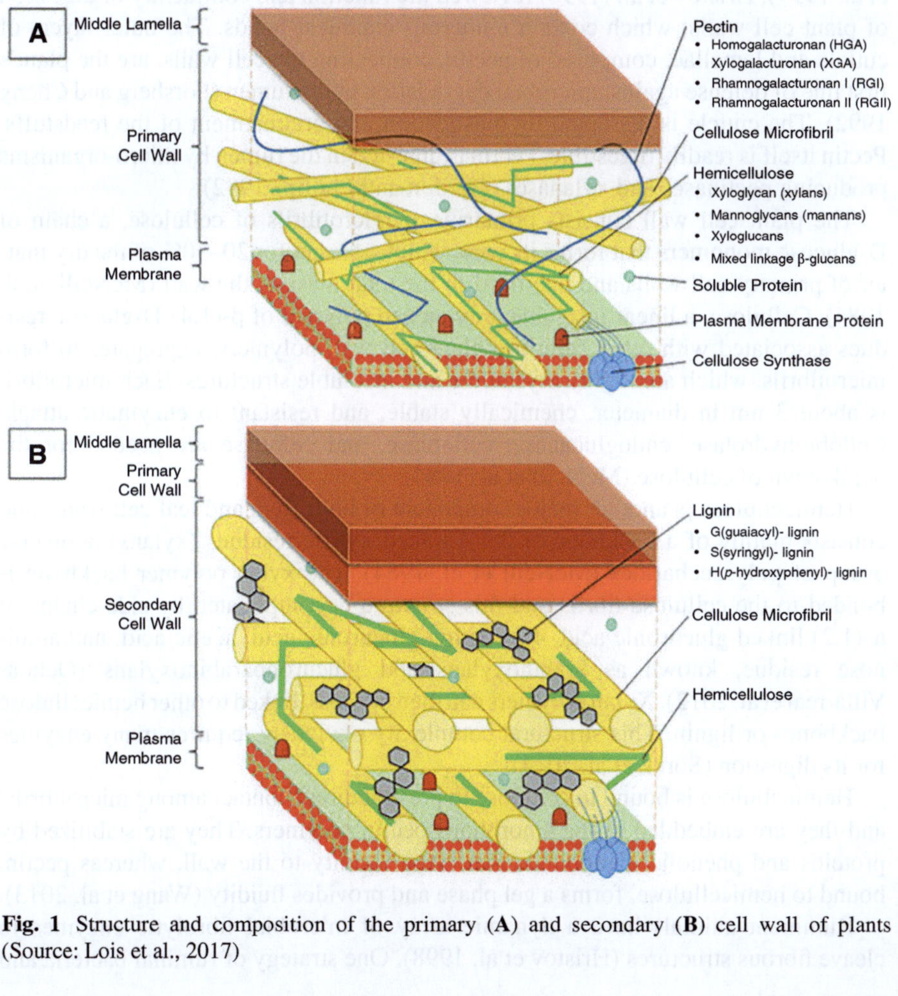

Fig. 1 Structure and composition of the primary (**A**) and secondary (**B**) cell wall of plants (Source: Lois et al., 2017)

lignin in between the polysaccharide molecules (Figure 1B). Furthermore, McManus and Bigham (1973) reported that the distribution pattern of lignin, for instance, rather than the total amount of lignin, can mask the potentially digestible cell walls of forages (Travis et al. 1997). Marked interactions exist among the cell wall thickness, lignification, and other anatomical characteristics of forages and their digestibility (Wilson 1993). Certain plant species have a secondary cell wall interior to the primary cell wall. This lends increased structural strength to the plant and forms a formidable barrier against microbial invasion. A plant cell wall is a highly active and differentiated network, formed by a mixture of cellulose, hemicelluloses, pectins, and, in some cases, proteins and phenolic compounds. The cell wall composition of plants on a dry matter basis is approximately 30% cellulose, 30% hemicellulose, and 35% of pectin, including 1–5% structural proteins.

The fiber portion of the cell wall makes up 300–800 g/kg of forage dry matter and can be a major source of nutritional energy for ruminants; unfortunately, only less than 50% of this fraction is readily digested and utilized by the animal (Hatfield et al. 1999). Hristov et al. (1998) reviewed the function and complexity of digestion of plant cell walls, which contain numerous chemical bonds. The outer layers of cuticle and lamellae, composed of pectin, connecting the cell walls, are the plant's first line of defense against microbial degradation in the rumen (Forsberg and Cheng 1992). The cuticle is disrupted by mastication and pretreatment of the feedstuffs. Pectin itself is readily digestible. Pectin is digested in the rumen by micro-organisms producing pectinases and xylanases (Gordon and Phillips 1992).

The plant cell wall consists primarily of microfibrils of cellulose, a chain of D-glucose monomers that forms its core, which account for 20–30% of the dry matter of primary cell walls and one-third of the total mass of the wall (McNeill et al. 1984). Cellulose, a linear insoluble unbranched polymer of β-(1,4)-D-glucose residues associated with other cellulose plant cell wall polymers, aggregates to form microfibrils, which are highly crystalline and insoluble structures. Each microfibril is about 3 nm in diameter, chemically stable, and resistant to enzymatic attack. Cellobiohydrolase, endoglucanase, cellobiase, and cellulase are needed for the breakdown of cellulose (McNeill et al. 1984).

Hemicellulose is another major component of both stem and leaf cell walls, and consists mainly of a backbone of β-1,4 linked xylose residues (xylans), a diverse group of polysaccharides (McNeill et al. 1984). The xylan polymer backbone is bonded to the cellulose fibrils and this structure is complicated by side-chains of α-(1,2) linked glucuronic acid, 4-*o*-methyl glucuronic acid, acetic acid, and arabinose residue, known as arabinoxylans and glucuronoarabinoxylans (Ochoa-Villarreal et al. 2012). Xylan polymers can then be cross-linked to other hemicellulose backbones or lignin. This structural complexity obviously requires many enzymes for its digestion (Sorek et al. 2014).

Hemicellulose is bound to cellulose to prevent direct contact among microfibrils, and they are embedded in the amorphous pectin polymers. They are stabilized by proteins and phenolic compounds that bring rigidity to the wall, whereas pectin, bound to hemicellulose, forms a gel phase and provides fluidity (Wang et al. 2013).

Ruminant animals have a dynamic array of microbial fibrolytic enzymes to cleave fibrous structures (Hristov et al. 1998). One strategy of ruminal bacteria and fungi to support rapid rates of fiber hydrolysis is the synthesis of large amounts of fibrolytic enzymes, particularly cellulase. This is predominantly the strategy of ruminal fungi, which also produce lower amounts of fibrolytic enzymes, but of high specific activity (Wood et al. 1986). Ruminal bacteria, on the other hand, utilize a strategy in which the enzyme activity is located predominantly at the cell surface, which facilitates adhesion and degradation of the fiber. Fibrous cells can be digested by bacteria only from the interior, because the middle-lamella primary wall region is indigestible (Wilson and Mertens 1995). This is consistent with the "inside-out theory of plant digestion" as described by Cheng et al. (1991).

The most readily digestible plant tissues are located inside the plant, and therefore, intact plants are digested slowly. For microorganisms to digest the plant cell contents, access is gained only via the stomata, and therefore, mechanical disruption of plant material, such as chewing or grinding, becomes important to improve microbial access (Cieslak et al. 2013).

The accessibility of the plant cell wall to ruminal micro-organisms is complex and is described by three components (Boon et al. 2005). The accessibility of a tissue particle is the first component. This is related to the size of the particle, with large particles having only outer cell walls available for fermentation. Hence, the slow initial rate of fermentation of these is often because of lignified and poorly degradable tissue (Engels and Schuurman 1992). The second and third components are accessibility of the plant cell wall and of its components by ruminal microorganisms. These components are related to structural factors such as cell wall thickness. For instance, sclerenchyma cells increase their cell wall thickness to such an extent that the lumen diameter of the cell becomes so limited that the space is sufficient only for one microbe at a time (Boon et al. 2005). Finally, the highly digestible cell wall contents can be encrusted by indigestible lignin, making it almost impossible for the microorganisms or even their enzymes to find access to such components. Major constraints to fiber digestion are the matrix interactions among wall biopolymers and the low substrate surface area, and limited penetration of the nonmotile cellulolytic microbes into the cell lumen (Weimer 1996). In addition to the chemical and architectural composition of fiber, the rumen environment is an important contributing factor to fiber digestion. It is mostly the ruminal pH that serves as a constraint. Weimer (1996) discussed microbial interactions as a second environmental factor of importance for fiber digestion. There is significant competition for nutrients between individual species of ruminal cellulolytic bacteria and other bacteria. All these limitations in degradation could theoretically be overcome by the addition of exogenous fibrolytic enzymes to complement the rumen microbial system.

3 Treatment of Poor-Quality Roughages

A significant amount of work has been done over the last 40 years at research and development level on methods and techniques to improve the value of fibrous feeds (Wanapat et al. 1996). High-fiber feeds or poor-quality roughages consist of many types, including crops residues, poor meadow or lucerne hay, some native grasses, edible shrubs, mature natural herbage, and stems, and certain by-products from farms and factories that are grazed or collected manually at a very advanced vegetation stage during the dry season (Chenost 1995). These usually make up the bulk – and sometimes the only component – of ruminant diets in developing countries with a warm climate. Optimizing the use of these poor-quality roughages requires the utilization of treatments that improve not only the quality of the roughage but also the digestive utilization of these feeds through the application of appropriate supplementation and feeding techniques (Wanapat et al. 1996).

Methods for the improvement of the digestive utilization of poor-quality roughages can be grouped as (i) chemical methods, which include treatments with NaOH, ammonia, urea, and urine (Chenost 1995); (ii) physical methods which involves processing such as grinding, pelleting, and ensiling (Hull et al. 1974; (iii) genetic methods such as breeding (Mittler and Blumwald 2010; and (iv) biological methods such as inoculants and enzymes (Beauchemin et al. 2003).

When choosing a method, a few factors should be considered, for example, economics, effectiveness, and safety aspects for the producer, as well as the type of livestock species, availability, and feasibility. Processing such as chopping, grinding, and pelleting increases surface area and density, resulting in less sorting and higher intakes (Adesogan 2003). There are risks associated with over-processing such as reduction of the effective fiber content and digestibility, heat-damaged proteins, dust, and dry matter losses. Although less effective than soaking in diluted NaOH, forages are currently sprayed with diluted NaOH without rinsing before feeding owing to polluted wastewater disposal issues. It does increase roughage digestibility and intake, but it is hazardous and can cause alkali burns. Ammonia is even less effective than NaOH, and is also hazardous and rather costly. Urea is relatively safer than ammonia, but has more variable effects, which could result in ammonia toxicity if consumed in large quantities. Urea can increase digestibility and is cost-effective, but collection and related health issues are major problems (Adesogan 2003). Fungal treatments can decrease the lignin effect, but they degrade desirable carbohydrates in the process (Adesogan 2003). Whereas exogenous enzyme treatments such as cellulase and xylanase tend to increase digestibility and intake (Adesogan 2003).

3.1 Exogenous Fibrolytic Enzymes

The use of exogenous enzymes as feed additives to improve the utilization of roughage has gained popularity in recent years (Giraldo et al. 2008a, b). For many years, however, researchers were discouraged from using enzymes to enhance the utilisation of roughages in the total mixed rations (TMRs). This was because of the perception that these enzymes would be rendered ineffective owing to the action of ruminal proteolysis (Adesogan 2005) or deactivation of the exogenous enzymes in the feed manufacturing process. However, these concerns have been disproved in recent studies, which showed that supplementing exogenous fibrolytic enzymes could improve feed utilization and thereby enhance productivity in beef and dairy cattle (Beauchemin et al. 2003) and even small stock such as sheep (Cruywagen and Goosen 2004). As indicated by Bhat (2000), many earlier research findings showed an improvement in feed digestibility and animal production using exogenous enzymes (Burroughs et al. 1960; Rust et al. 1965).

During ruminant feeding, increasing the roughage proportion in the diet will increase mean retention time, which will lead to higher methane production and a decreased intake (Allen 1996), which might have a negative effect on performance (Gibb and McAllister 1999). Thus, supplementation with fibrolytic enzymes might have a positive role by improving digestibility (Giraldo et al. 2008a, b; Bala et al. 2009) and intake (Bala et al. 2009) of the roughage proportion, which would decrease the retention time. Cellulases, hemicellulases, xylanase, and even pectinases are used in ruminant feed biotechnology to improve feed utilization. The majority of research on fibrolytic enzyme additives in ruminants has been conducted on beef and dairy cattle. However, only a few examples of such improvements in animal performance are reported, which include enhancement in milk yield, fat-corrected milk, energy-corrected milk (Mohamed et al. 2013), live weight gain (Beauchemin et al. 1995), feed intake and digestibility of dry matter and NDF in cattle (Feng et al. 1996). Nonetheless, negative effects were also shown in some of these early studies (Theurer et al. 1963; Perry et al. 1966). However, great strides have been made in the understanding of enzymes and their application, as is evident in the range of exogenous fibrolytic enzymes that are available commercially.

The use of exogenous fibrolytic enzymes has shown positive effects in terms of modulation of rumen fermentation and improvement of fiber degradation (Adesogan et al. 2007; Krueger et al. 2008a, b). Cellulase and xylanase in particular, which were introduced to ruminant diets, exhibited positive results which includes: enhanced dry matter intake (Beauchemin et al. 1999), increased digestibility (Krueger et al. 2008a, b), improved live weight gain of beef cattle (Beauchemin et al. 1999) and milk production of dairy cows (Adesogan et al. 2007). Further improvement was achieved through the use of proteolytic enzymes in combination with fibrolytic enzymes (Gado et al. 2009), although the results need to be verified for substrates with different characteristics and enzymes from various sources. The intended benefit of the addition of a proteolytic enzyme is that this enzyme would degrade the cell wall protein and improve access of ruminal microbiota to more digestible substrates (Colombatto and Beauchemin 2009). Elghandour et al. (2013) reported improvements in degradability and fermentation parameters in vitro with an enzyme cocktail containing endoglucanase, α-amylase, xylanase, and protease. Gado et al. (2009) confirmed these findings after conducting in vivo studies using the same enzyme cocktail. Although research on exogenous fibrolytic enzymes began in the early 1950, their production then became easy and economical in the early 1980s owing to the advances in fermentation technology and biotechnology. Exogenous fibrolytic enzymes revealed their biotechnological potential in various industries, including animal nutrition (Bhat 2000). Currently, several fibrolytic enzyme supplements are available commercially for use in improving fiber degradation (Adesogan 2005). The main aim of the inclusion of these enzymes is to increase the fiber digestibility of the diets, with subsequent improvements in feed intake and animal production. Preparations of enzymes that degrade cell walls (cellulases and xylanases) have the potential to hydrolyse forage fibre (Feng et al. 1996). Hristov et al. (1998) defined the mode of action of exogenous enzymes as proteins that catalyze chemical reactions in biological systems. In the context of animal feeds, exogenous enzymes catalyze the degradation reactions of feedstuffs to release nutrients for utilization by the micro-organisms or the host animal itself.

Exogenous fibrolytic enzyme additives, primarily xylanases and cellulases, are concentrated extracts resulting from bacterial or fungal fermentation that have specific enzymatic activities (Beauchemin et al. 2003). The addition of commercial exogenous fibrolytic enzymes such as cellulase and xylanase offers opportunities to increase the proportion of the roughage component in the TMR or to use very low-quality roughage during feed shortage or high feed prices. These enzymes normally increase digestion without negative effects on the intake, though improved performance is not always the case owing to a variety of interacting factors such as the type of enzyme, source, and concentration (El-Kady et al. 2006). The successful use of enzymes depends on their stability in the feed, their ability to hydrolyze the plant cell wall components, and the ability of the animal to utilize the resultant products efficiently (Bhat 2000). Their effectiveness varies with the substrate composition, the method of enzyme application, and the component of the diet to which the enzyme is added (Wallace et al. 2001).

Cellulase is responsible for the hydrolysis of cellulose. The main enzymes involved in hydrolysis are endocellulase, exocellulase, and β-glucosidase, whereas the main enzymes involved in degrading xylan core polymer to soluble sugars are xylanase, β-1,4 xylosidase, and endoxylanase. There is evidence that the mode of action of exogenous enzymes in ruminants is a combination of pre- and post-feeding effects (Colombatto and Beauchemin 2003). These prefeeding effects include an enzyme–substrate preincubation interaction period. In a review paper by Alvarez et al. (2009), it was reported that several researchers had suggested that preincubation of the diet with the enzyme was important (Elwakeel et al. 2007; Krueger et al. 2008a, b). Exogenous enzymes are more stable when applied to the feed prior to ingestion, as opposed to infusion into the rumen (Beauchemin et al. 2003). Application of the enzymes to the feed allows for absorption and binding time to the substrate, enhances the binding of the enzyme to the substrate, making it more resistant to proteolytic breakdown in the rumen, and thereby prolonging the resident time of the feed and enzymes (Beauchemin et al. 2003). The resultant stable enzyme–feed complex could then degrade the plant tissue in the rumen (Kung et al. 2000) and provide a slow-release mechanism for the enzymes there.

No improvements were observed in degradation (Kopecny et al. 1987) when enzymes were infused directly into the rumen instead of applying it to the feed, which serves as further justification for allowing a preincubation interaction period. In Giraldo et al. (2008a, b), the authors found that direct-fed fibrolytic enzymes affected fibrolytic activity in the rumen of sheep positively and increased the growth of cellulolytic bacteria without a pre-feeding substrate enzyme interaction period.

They reported an increase in the ruminally insoluble potential degradable fraction of grass hay DM, and in its fractional rate of degradation. However, enzyme supplementation did not affect diet digestibility, even though molar proportions of propionate were greater and acetate to propionate ratio was lower.

Alvarez et al. (2009) reported that the reducing sugars released because of the addition of fibrolytic enzyme provided energy, which should lead to rapid microbial growth, which would increase ruminal bacteria numbers and could lead to increased microbial colonization of the feed particles. Giraldo et al. (2008a, b) suggested an alteration in the fiber structure owing to enzyme effects. An altered fiber structure in combination with increased colonization would shorten the lag time (Yang et al. 1999), and the access of the microbes to the potentially fermentable fiber by enzymes acting on the structures of plant cell walls would be enhanced (Sutton et al. 2003; Elwakeel et al. 2007). The beneficial activities from adding exogenous fibrolytic enzymes to a diet are related to their ability to enhance the initial degradation of a plant's structural carbohydrates and complement normal enzymatic activities associated with ruminal microorganisms (Dawson and Tricarico 1999).

3.2 *Exogenous Fibrolytic Enzymes in the Nutrition of Ruminants*

Although exogenous fibrolytic enzymes present one way of improving fibre digestibility (Johnston 2000) their supplementation in the diet of small ruminants has received relatively little attention compared with large ruminants (Giraldo et al. 2008a, b).

Positive effects have been reported of supplementing diets of large ruminants with exogenous fibrolytic enzyme. These results were obtained with pre-treatment of the test feeds with the enzymes, but direct-fed enzymes also showed positive effects. These effects included increased NDF digestion in bulls (Murillo et al. 2000) and dairy cattle (Lopez-Soto et al. 2000), increased average daily weight gain in steers (Ware et al. 2002), improved in vitro (Colombatto and Beauchemin 2003) and in vivo (Schingoethe et al. 1999) cell wall digestibility for some exogenous fibrolytic enzyme preparations. In addition, other studies indicated an increase in milk production of lactating dairy cows (Kung et al. 2000; Yang et al. 2000), improvement in the energy balance of transition cows (DeFrain et al. 2005), increase in forage utilization and production efficiency, and reduced nutrient excretion (Beauchemin et al. 2003). Increased feed intake was recorded in some cases. This was attributed to increased palatability associated with the preingestive release of sugars during fiber hydrolysis and postingestive increases in digestion rate, extent of digestion, and passage rate (Adesogan 2005).

Most of the positive effects of enzyme-supplemented feeds on production responses were probably because of ruminal effects (Adesogan 2005). Supplementation increases the hydrolytic capacity of the rumen (indirectly reducing gut fill, increasing passage rate and intake) by increasing bacterial attachment, stimulating rumen microbial populations, and causing synergistic effects with hydrolysis of ruminal microorganisms with the net effect being increased enzyme activity in the rumen, which increases the digestibility of the feed (Beauchemin et al. 2003).

In semiarid and arid regions, sheep and goats are important being because of their adaptation to these climates (Bala et al. 2009). Studies using goats to ascertain the effects of exogenous fibrolytic enzymes have been limited and the responses were poor because of the superior ability of goats to utilize fiber compared with that of large ruminants (Bala et al. 2009). In addition, there is a dearth of information on the effects of exogenous fibrolytic enzymes on ruminal fermentation in small ruminants (Pinos-Rodriguez et al. 2002). Yang et al. (2000) could not ascertain any effects of fibrolytic enzymes fed to goats, whereas Bala et al. (2009) ascertained positive results on DM, organic matter (OM), NDF, acid detergent fiber (ADF), crude protein (CP), and total carbohydrate digestibility, and increased milk yield in the third quarter of lactation, along with a decrease in feed intake of up to 7%. Researchers observed positive results such as improved ruminal activity (Giraldo et al. 2008a, b), increased body weight gain, and feed conversion ratios in sheep fed on enzyme treated diet (Cruywagen and Goosen 2004). Booyse and Hassen (2018) (unpublished data) found improvement in in vitro gas production of poor-quality hay during the early hours of fermentation, with an associated increase in acetate, propionate, butyrate, and total ruminal volatile fatty acid production. However, no beneficial effect of fibrolytic enzyme addition was observed during their in vivo evaluation in digestibility, intake and, as well as growth performance of South African Mutton Merino lambs, fed a high forage TMR. In their study, the TMR had oil or -rumen-protected fat and this might partly explain the lack of response in digestibility and growth performance of the sheep. Thus, small ruminants have been considered in a limited number of studies with regard to their response to fibrolytic enzyme treatments (Marquardt and Brufau 1997).

3.3 *Exogenous Fibrolytic Enzymes Research Techniques*

The abundance of research on exogenous fibrolytic enzyme application appears to have been done using in vitro and in situ studies. In vitro determination of the correct application is important. There are several factors that one needs to consider to allow direct comparisons among studies in which different enzyme products are used. The temperature and pH at which enzyme activities are assayed are most important when evaluating enzymes for specific purposes, for example, feeding direct-fed or ensiling. The recommended conditions for conducting an assay are pH 6.0–6.5 at 39 °C, which reflects that of ruminal conditions. Because enzyme activity is expressed as an initial rate of reaction, it is important to ensure that the enzyme is at the maximum kinetic velocity for the duration of the assay. This can be done by making sure that the substrate concentration is well above that of the enzyme. Incubation time should not be too short and the number of samples that can be run at the same time should not be reduced. If the incubation time is also too long, the chances of obtaining misleading results are increased (Colombatto and Beauchemin 2003).

The degradation kinetics of enzyme-treated feeds has been described mainly using in situ techniques. However, these methodologies are limited in the number of samples that can be analyzed at one time, and they assume that all feed losses from the bags have been or are degradable. There is therefore a need to develop suitable in vitro bioassays (for example in vitro gas production) to provide an alternative to such in situ trials, especially when screening large numbers of samples or treatments (Colombatto and Beauchemin 2003).

In vitro gas production is a useful technique for measuring the effects of the exogenous enzyme treatment of forages. In this technique, head space gas production can be measured throughout incubation. Eun and Beauchemin (2007) evaluated 13 endoglucanases and 10 xylanases in this manner, and were able to show increased gas production and organic matter digestibility (OMD) at 18 hours with many of these enzymes when applied to lucerne hay. Based on these initial screenings, two superior enzymes of each category were evaluated, also in combination. The authors found that the enzymes were effective in improving gas production and organic matter digestibility, but there was no associative effect with the combination of the enzymes. Gas production-based techniques could be an alternative to in situ techniques as they account for insoluble and soluble fractions. However, because gas production profiles are not necessarily related directly to substrate degradation, determination of the undegraded residue was recommended (Colombatto and Beauchemin 2003).

4 Sustainable Animal Production

The production of sufficient quantities of animal protein to feed an ever-expanding world population has become one of the primary objectives and concerns of our time. The challenge to produce enough animal protein is complicated because it has to be produced in such a manner as to sustain our planet. To achieve this goal, global production must curtail greenhouse gas emissions, deforestation, loss of biodiversity and excretion of excessive nitrogen and phosphorous to the environment. Therefore, humankind needs to function within the critical confines of the planet's biosphere and boundaries. This requires continuous use of innovative and research-based solutions that reduce the amount of resources – land, water, energy, and feed – needed to produce animal products. In recent times, nations have increased their efforts to combat climate change. The United Nations estimated that the global population would increase from 7.3 billion in 2015 to 9.7 billion by 2050 (Elghandour et al. 2013). This estimation was made by its medium-variant projection. This figure is stated with some uncertainty, because factors such as economic growth, educational progress, national health, and family planning can alter population growth.

Even minor differences in fertility rates could have a major effect on it. In addition, over the last couple of decades, per capita income has increased in most nations, which has resulted in shifts in consumption patterns towards high-quality protein from domestic and game animals. Therefore, the demand for animal protein such as meat, eggs, fish, and milk has increased. In some more affluent nations, overconsumption of protein has resulted in imbalanced diets. Consumption of excessive amounts of red meat and processed meat may result in health risks such as obesity, diabetes, cardiovascular disease, and even cancer (Worldwatch 2016). The global middle class is projected to increase to an estimated 5 billion by 2030. This is a significant expansion from 1.8 billion in 2009, and most of this growth is projected to occur in Asia (Kharas 2010). The Food and Agriculture Organization (FAO 2009) of the UN studied the correlation between protein intake and increasing affluence. They found that per capita protein intake increased in a linear fashion when annual income ranged below 10,000 US dollars. Furthermore, an ever-increasing proportion of the global population is moving from rural to urban areas. Urbanized populations tend to consume a greater range of dietary components compared with rural populations, which are defined by a higher intake of meat, eggs, milk, and fish (Forbes 2016). Hence, the goal should not be confined to producing adequate food for the ever-growing world population but extended to provide a balanced diet with high-quality protein for everyone.

4.1 Ruminal Digestion and Methane Production

Feed ingredients are digested in the rumen through symbiosis between the host animal and microbes (McDonald et al. 2011). Microorganisms such as ciliated protozoans play a critical role in fibre fermentation by improving the attachment of other microorganisms to feed particles. They do this by engulfing feed particles that enter the anaerobic rumen themselves, whereas others digest rumen-degradable particles through secretions of fibrolytic enzymes (Krause et al. 2003). The rumen is a complicated environment consisting of various complementary and opposing relationships among microorganisms. Ruminal digestion is further complicated by the diversity of rumen microorganisms, which consist of a wide spectrum of bacteria, protozoa, fungi, and methanogenic microorganisms (Cieslak et al. 2013). The presence of fiber-fermenting bacteria that produce hydrogen (H_2) is related directly to methanogens, because of hydrogen transfer between these microbes (Morgavi et al. 2010). The animal then has the primary products of microbial fermentation available for utilization (Krause et al. 2003). These include volatile fatty acids, microbial protein, carbon dioxide (CO_2), and methane (CH_4) (Murray et al. 1976). The production of CH_4 takes place through the exchange of metabolites between H-synthesizing and H-utilizing microorganisms (Kobayashi 2010). This continuous exchange of H enables constant fiber fermentation in the rumen (Ushida et al. 1997).

Methane production accounts for a significant amount of loss of gross energy ingested, depending on the type of ingested material (Hristov et al. 2013). However, a certain degree of CH_4 production is required to maintain a healthy rumen environment as it restricts hydrogen build-up, which prevents the inhibition of dehydrogenase required in the oxidation of reduced co-factors (Martin et al. 2009). Following the reduction of cofactors such as NADH (nicotinamide adenine dinucleotide – hydrogen), NADPH (nicotinamide adenine dinucleotide phosphate– hydrogen), and FADH (flavin adenine dinucleotide hydrogenate), methanogens can utilize the hydrogen and reduce CO_2, which results in the production of CH_4. The associated loss of energy from this process can be exacerbated when low-quality diets are fed that lack sufficient essential nutrients for healthy ruminal microbial growth (Leng 1991).

The production of meat and milk products from ruminants for human consumption is increasingly seen as one of the main contributing factors to harmful greenhouse gases (De Vries and De Boer 2010). On the other hand, the global population is projected to reach more than 9.7 billion people by 2050 with a concomitant 60% increase in demand for food and livestock products (Alexandratos and Bruinsma 2012). As a result of this rising demand, livestock production is expected to double by 2050 (Garnett 2009). This necessitates more research on the efficient utilization of feed resources in combination with mitigation strategies aimed at reducing the ever-increasing harmful greenhouse gases emitted from the livestock sector. Sustainable animal production, guided by increased efficiency of resource utilization and reduced environmental impact, is seen as the solution. Two mechanisms of such a solution are better utilization of feed ingredients through improvement in digestibility and an improved digestive environment (Binder 2016). More recent studies indicated that sustainable animal nutrition would not be possible without modern feed additives under intensive production system (Alexandratos and Bruinsma 2012). Feed additives such as enzymes have the potential to lower the amount of resources (land, feed, water, and energy) used in livestock production, thereby leading to the production of healthy nutritious food for a growing world population (De Vries and De Boer 2010).

Methane, along with carbon dioxide and nitrous oxide, is an important greenhouse gas, which contributes to the global warming (Steinfeld et al. 2006). In livestock production, CH_4 is produced in anaerobic conditions in the rumen by enteric methanogenic bacteria that use carbon dioxide and hydrogen as building blocks (Leng 2008). Various factors influence the production of CH_4 in the rumen. The effects of diet and substrate degradability on the mitigation of ruminal CH_4 are factors that could be altered (Singh et al. 2005; Smith et al. 2005). The production of CH_4 is desirable from a fermentative standpoint. However, it requires a significant amount of feed energy to produce methane. Cattle spend between 2% and 12% of their gross energy on the synthesis of this greenhouse gas (Beauchemin et al. 2009), but depending on the type of ingested material, Hristov et al. (2013) found methane production accounting for 7–17% of the loss of gross energy ingested by ruminants. Therefore, nutritional interventions that contribute to the improvement of the efficiency of nutrient utilization, reduction of energy losses, and mitigation of enteric CH_4 production should be implemented as favorable strategies (Monteny and Chadwick 2006; Beauchemin et al. 2008). However, it is also crucial to introduce economical and low-risk mitigation strategies in a responsible manner to ensure that these strategies are adopted by livestock producers and that the integrity of the food supply chain remains intact.

4.2 Constraints to Fermentation of Fibrous Feed and Its Utilization

The outer layer of plant cell walls consists of pectin, cuticle, and epicuticular waxes, which enable the plant to defend itself against dehydration and the invasion of phytopathogens. However, these protective layers of forages, grains, and legumes may act as obstructions to feed degradation by the ruminal microbiota in the rumen (Selinger et al. 1996). These obstructions restrict microbial attachment to plant cell walls, thereby limiting ruminal fermentation. The infiltration of plant cell walls by the microbiota usually occurs through mechanical breakdown, namely, chewing, cutting, grinding, and pelleting at the stomata and lenticels. The extent of microbial establishment and the mechanisms of attachment differ among microbial species in the rumen (Varga and Kolver 1997). The attachment is critical for effective fibrous feed fermentation and requires a well-balanced and adapted rumen microbial environment (McAllister et al. 1994). The microbial population binds to the cell walls in various ways, including specialized binding proteins and receptors and physicochemical forces such as Van der Waals forces (McAllister et al. 1994).

Breakdown of plant cell walls in the rumen takes place mostly because of the presence of fibrolytic bacteria such as *Fibrobacter succinogenes, Ruminococcus flavefaciens*, and *Ruminococcus albus,* which secrete fibrolytic enzymes with specific activities (Weimer 1996; Varga and Kolver 1997). During this process, the excreted cellulolytic enzymes adhere to the substrate, preventing other microorganisms and their enzymes from having access to the site of hydrolysis. Therefore, rumen cellulolytic bacteria can utilize the products of cellulolytic hydrolysis first (McAllister et al. 1994). Less is known about the exact role of fungi and protozoa compared with bacteria. Fungi infiltrate the cell wall of lignified tissue. Fungal enzymes possess a broad range of activities, making fungal cellulases and xylanases the most active fibrolytic enzymes (Selinger et al. 1996). Similarly, protozoa contain all of the primary enzyme activities and therefore have a great capacity to digest plant cell walls (Selinger et al. 1996). McDonald et al. (2011) indicated that the ruminal microbial population works in a complementary manner in the digestion of fibrous components by secreting a broad range of enzymes with multiple activities.

Various factors can lead to the restriction of fibrous feed degradation in the rumen. These include

1. The physical and chemical arrangement of plant components that prevent microbial colonization (McAllister et al. 1994)
2. The balance among species of microorganisms in the rumen, the interactions between these species, the type of fibrolytic enzymes secreted, and their relative

concentrations as well as the manner of colonization of each species (Varga and Kolver 1997)
3. Secreted microbial fibrolytic enzymes, which affect colonization and hydrolysis of other microbial species (McAllister et al. 1994)
4. The particle size of the substrate, which determines the area available for initial microbial colonization (Buxton and Redfearn 1997)
5. The degree of lignin, because it acts as a physical barrier to microbial colonization, leading to many substrates that can be digested only from the interior (Fisher et al. 1995)
6. The pH of the rumen, which affects the rumen microbial environment (Plaizier et al. 2009)

A decrease in ruminal pH can be expected when feeding more grain and less roughage. The production of less sodium bicarbonate occurs because the need for mastication and rumination is lowered. This shift inevitably results in a reduction of ruminal pH, which can lead to a decline in cellulolytic microorganisms and ultimately fiber digestion. It is critical that the diet should contain sufficient fiber and the correct particle size to encourage rumination and combat low ruminal pH (Grant 1997). Differences observed in the mastication actions of animals may affect rumination and contractions in the digestive tract, which can influence the utilization of fibrous substrates (Firkins et al. 1998).

4.3 *Exogenous Enzymes in Ruminant Nutrition*

Digestive enzymes are of particular importance to nutritionists as they play a critical role in the breakdown of nutrients in the gastrointestinal tract of animals (McDonald et al. 2011). In animal nutrition, enzymes can be further divided into two groups, namely, endogenous (originating from the animal) and exogenous (originating from microorganisms). These enzymes thus increase the rate of degradation of feed substrates that can be absorbed by the animal itself or the microbial population within the animal (Krueger et al. 2008a, b).

Suitable exogenous enzymes can be added to feed substrates to increase their nutritive value by improving the efficiency of digestion and absorption. Various types are available with different modes of action, such as fibrolytic enzymes, proteolytic enzymes, amylolytic enzymes, and phytases, which act on the fiber, protein, starch, and unavailable phosphorous fraction of feedstuffs, respectively (Bedford and Partridge 2011).

Investigation into the use of exogenous fibrolytic enzymes in ruminants began in the early 1950s. This was based on their potential to convert lignocellulose to glucose and other soluble sugars. Lignocellulose is the most abundant and renewable source of energy on earth but is slowly degradable. After advances in fermentation technology and biotechnology, the industrial production of exogenous fibrolytic enzymes was initiated in the 1980s (Bhat 2000). In animal nutrition, the use of feed enzymes showed the potential to enhance the efficiency with which animals utilized raw materials (Beauchemin et al. 2003). Ruminal digestion and absorption of plant cell walls are inefficient. According to Van Soest (1994), less than 65% of the potential nutritive value of plant cell walls is degraded in the rumen.

Commercial enzymes, commonly referred to as cellulases or xylanases, rarely consist of pure enzyme products, and usually contain secondary enzyme activities such as amylases, proteases, esterases, and pectinases (McAllister et al. 2001). This is advantageous as a wide spectrum of enzymes is required to degrade the structural carbohydrates in plant cell walls (Morgavi et al. 2000a). However, the identification of pure enzymes responsible for positive interaction with feed digestion becomes a challenge. Common practice is to include enzymes that are suitable for a range of feed types (Beauchemin et al. 2003). This always leads to highly variable results when enzymes are included in ruminant formulations. Investigation of the effects of pure enzyme products on specific feedstuffs can enable more precise formulation. The addition of exogenous enzymes to suitable and commonly used feedstuffs would ultimately result in more reliable performance. Fibrolytic enzymes could result in increased nutrient digestibility. These nutrients can then be utilized by the microbial population in the rumen, facilitating more rapid microbial growth, and increasing microbial colonization of feed substrate particles (Alvarez et al. 2009). Giraldo et al. (2008a, b) observed a shift in the fiber structure because of the enzymatic effects. This, together with increased feed colonization, constituted the lag time before the start of microbial digestion (Yang et al. 1999; Elwakeel et al. 2007). There is a synergistic effect between certain exogenous fibrolytic enzymes and microbial enzymes produced in the rumen.

Proteolytic enzymes are involved in the utilization of proteins and therefore make a direct contribution to the growth of an animal. Studies were conducted on the addition of proteolytic enzymes in combination with fibrolytic enzymes. The assumption was that proteolytic enzymes would degrade the soluble nitrogen interlocked with the cell wall fiber, thereby improving the availability of more digestible nutrients to rumen microorganisms (Colombatto and Beauchemin 2009). Enzyme cocktails containing endoglucanase, xylanase, α-amylase, and protease have shown increased digestibility and ruminal fermentation during in vitro (Elghandour et al. 2013) and in vivo (Gado et al. 2009) evaluation studies. The addition of protease to whole-corn plants after harvest increased the proteolytic stages of the ensiling process, resulting in greater ruminal in vitro starch degradation (Young et al. 2017). The addition of exogenous protease enzyme to low-quality roughage may increase digestibility and reduce CH_4 per organic matter digested (Beauchemin et al. 2008; Grainger and Beauchemin 2011), but the results varied substantially. An in vivo screening by McGinn et al. (2004) did not reveal a reduction in CH_4 per organic matter digested.

4.4 Factors Affecting Enzyme Efficacy in Ruminants

The effectiveness of exogenous enzymes is measured by their ability to improve the digestibility of feed and unlock feed nutrients to an extent that they can be incorporated by the animal itself or via the animal's microbial population. Thus, exogenous enzymes should supplement the enzymatic activities that are limiting digestion (Morgavi et al. 2000b). Various factors contribute to the efficacy of exogenous enzymes of which enzyme–substrate preparation and the interaction period prior to feeding play a major role.

4.5 Enzyme–Substrate Preparation and Interaction Period Prior to Feeding

The high moisture content of the feed substrate can increase the hydrolysis of carbohydrates from complex polymers (Wallace et al. 2001). The subsequent solubilization of ADF and NDF can cause the rapid spread of microorganisms and reduction of the lag time needed for microbial establishment (Devillard et al. 2004). Therefore, the application of liquid enzyme treatment to a dry substrate is recommended. Furthermore, applying a preincubation time of enzymes and substrate together enables absorption and binding of the enzyme to the substrate. Enzymes that interacted with the substrate prior to consumption appeared to have higher efficacy in vivo. This was probably because of its higher resistance to proteolytic inactivation in the rumen, resulting from a protective enzyme–substrate complex that had formed (Kung et al. 2000). Colombatto et al. (2007) recommended an enzyme–substrate interaction time of 12–36 hours before consumption. In addition, increasing the proportion of substrate to which the enzyme was added improved its viability in the rumen. It was found that adding enzymes to the macro raw materials that make up 45% of the ration was more effective than including them in the premix, which comprised 0.2% of the diet (Bowman et al. 2002).

4.6 Enteric Methane Mitigation Strategies and the Role of Enzyme to Reduce Methane Emission Intensity

Various factors contribute to enteric CH_4 production, namely, the availability and composition of ingested feed (Beauchemin et al. 2008), the ratio of acetate to propionate in the rumen (McAllister et al. 1996), the types of carbohydrates after fermentation (Santoso et al. 2003), NDF and ADF concentration (Hindrichsen et al. 2003), and the characteristics of the forage that is consumed (Arthington and Brown 2005).

Alterations in the ration by lowering the proportion of fiber in exchange for concentrate should reduce CH_4 production per unit DM consumed (Lovett et al. 2003). In a study by Agle et al. (2010), a dairy cow ration with 50% concentrate produced 44% higher methane compared with a diet with 70% concentrate. Ruminants consuming feed with a higher proportion of concentrates such as grain or starch generally emitted lower CH_4. However, multiple livestock production systems in developing countries cannot use concentrates as a methane mitigation strategy because of the cost implications and the lack of concentrates. Many researchers concluded that increasing the quality of forage fermentation (for example increasing the sugar content of grasses) could result in the reduction of enteric CH_4 emissions (Boland et al. 2009; McGeough et al. 2010). In contrast, improving the fermentation of poor-quality forage with fibrolytic exogenous enzymes often resulted in higher ruminal fermentation, characterized by higher production of acetate and total VFAs. The improvement in fermentation and the digestibility of fiber by ruminants under extensive livestock production seems to increase methane emission per unit of dry matter digested, but it will indirectly reduce the intensity of methane emission per unit of products because of improved VFA production, efficient utilization of the feed and improvement in productivity of herds in these production systems that rely on roughage as the bulk of the feed source. In addition, simultaneous use of other additives that serve as a hydrogen sink (such as nitrate and oil) would further reduce the enteric methane production per unit of feed consumed and per unit of feed digested (Gemeda et al. 2018, unpublished data).

Acknowledgments The authors acknowledge using a figure prepared by Loix et al (2017) in this chapter for illustration purpose and they are grateful for both the authors and publishers for making the material available under Public License, Attribution 4.0 International (CC BY 4.0). https://creativecommons.org/licenses/by/4.0/#:~:text=Attribution%204.0%20International,CC%20BY%204.0

References

Adesogan AT (2003) Improving the intake and digestibility of poor quality forages. In: Proceedings of the conference titled advances in feeding and management of dairy cattle. University of La Molina, Lima

Adesogan AT (2005) Improving forage quality and animal performance with fibrolytic enzymes. Florida ruminant nutrition symposium, 2005

Adesogan AT, Ma ZX, Romero JJ et al (2007) Improving cell wall digestion and animal performance with fibrolytic enzymes. J Anim Sci 92:1317–1330

Agle M, Hristov AN, Zaman S et al (2010) Effects of ruminally degradable protein on rumen fermentation and ammonia losses from manure in dairy cows. J Dairy Sci 93:1625–1637

Alexandratos N, Bruinsma J (2012) World agriculture towards 2030/2050: the 2012 revision. ESA working paper no. 12-03. Food and Agriculture Organization, Rome

Allen MS (1996) Physical constraints on voluntary intake of forages by ruminants. J Anim Sci 74:3063–3075

Alvarez G, Pinos-Rodriguez JM, Herrera JG et al (2009) Effects of exogenous fibrolytic enzymes on ruminal digestibility in steers fed high fibre rations. Livest Sci 121:150–154

Arthington JD, Brown WF (2005) Estimation of feeding value of four tropical forage species at two stages of maturity. J Anim Sci 83:1726–1731

Bala P, Malik R, Srinivas B (2009) Effect of fortifying concentrate supplement with fibrolytic enzymes on nutrient utilization, milk yield and composition in lactating goats. J Anim Sci 80:265–272

Beauchemin KA, Rode LM, Sewalt VJH (1995) Fibrolytic enzymes increase fibre digestibility and growth rate of steers fed dry forages. Can J Anim Sci 75:641–644

Beauchemin KA, Yang WZ, Rode LM (1999) Effects of grain source and enzyme additive on site and extent of nutrient digestion in dairy cows. J Dairy Sci 82:378–390

Beauchemin KA, Colombatto MDP et al (2003) Use of exogenous fibrolytic enzymes to improve feed utilization by ruminants. J Anim Sci 81:37–47

Beauchemin KA, Kreuzer M, O'Mara F et al (2008) Nutritional management for enteric methane abatement: a review. Aust J Exper Agric 48:21–27

Beauchemin KA, McAllister TA, McGinn SM (2009) Dietary mitigation of enteric methane from cattle. Perspect Agric Vet Sci Nutr Nat Resour 4:1–18

Bedford MR, Partridge GG (2011) Enzymes in farm animal nutrition, 2nd edn. CABI, Wallingford, pp 12–54

Bhat MK (2000) Cellulases and related enzymes in biotechnology. Biotechnol Adv 18:355–383

Binder M (2016) Life cycle thinking in animal production. AMINO News. Special edition. 2016. Sustainability

Birhan M, Adugna T (2014) Livestock feed resources assessment, constraints and improvement strategies in Ethiopia. Middle-East J Sci Res 21:616–622

Boland TM, Hart KJ, Pierce KM et al (2009) The effect of pre-grazing herbage mass on growth rate and methane emissions of grazing beef cattle. J Dairy Sci 92:339–343

Boon EJMC, Engels FM, Struik PC (2005) Stem characteristics of two forage maize (Zea mays L.) cultivars varying in whole plant digestibility. II. Relation between in vitro rumen fermentation characteristics and anatomical and chemical features within a single internode. Neth J Agric Sci 53:87–109

Bowman GR, Beauchemin KA, Shelford JA (2002) The proportion of the diet to which fibrolytic enzymes are added affects nutrient digestion by lactating dairy cows. J Dairy Sci 85:3420–3429

Burroughs W, Woods W, Ewing SA (1960) Enzyme additions to fattening cattle rations. J Anim Sci 19:458–464

Buxton DR, Redfearn DD (1997) Plant limitations to fibre digestion and utilization. J Nutr 127:814–818

Cheng KJ, Forsberg CW, Minato H et al (1991) Microbial ecology and physiology of feed degradation within the rumen. In: Tsuda T, Sasaki Y, Kawahima R (eds) Physiological aspects of digestion and metabolism in ruminants, Proceedings of the seventh international symposium on ruminant physiology, pp 595–624

Chenost M (1995) Optimizing the use of poor quality roughages through treatments and supplementation in warm climate countries with particular emphasis on urea treatment. First electronic conference on tropical feeds. FAO, Rome

Cieslak A, Szumacher-Strabel M, Stochmal A (2013) Plant components with specific activities against rumen methanogens. Anim Sci 7:253–265

Colombatto D, Beauchemin KA (2003) A proposed methodology to standardize the determination of enzymatic activities present in enzyme additives used in ruminant diets. Can J Anim Sci 83:559–568

Colombatto D, Beauchemin KA (2009) A protease additive increases fermentation of alfalfa diets by mixed ruminal microorganisms *in vitro*. J Anim Sci 87:1097–1105

Colombatto D, Mould FL, Bhat MK et al (2007) Influence of exogenous fibrolytic enzyme level and incubation pH on the in vitro ruminal fermentation of alfalfa stems. Anim Feed Sci Technol 137:150–162

Cruywagen CW, Goosen L (2004) Effect of an exogenous fibrolytic enzyme on growth rate, feed intake and feed conversion ratio in growing lambs. S Afr J Anim Sci 34:71–73

Dawson KA, Tricarico JM (1999) The use of exogenous fibrolytic enzymes to enhance microbial activities in the rumen and performance of ruminant animals. In: Lyons TP, Jacques KA (eds) Biotechnology in the feed industry, Proceedings of the 15th annual symposium, p 303

De Vries M, De Boer IJM (2010) Comparing environmental impacts for livestock products: a review of life cycle assessments. J Livest Sci 128:1–11

DeFrain JM, Hippen AR, Kalscheur K et al (2005) Feeding alpha amylase improves the glycemic status and performance of transition dairy cows. J Dairy Sci 88:4405–4413

Devillard E, Goodheart DB, Karnati SKR et al (2004) *Ruminococcus albus* 8 mutants defective in cellulose degradation are deficient in two processive endocellulases Cel48A and Cel9B, both of which possess a novel modular architecture. J Bacteriol 186:136–145

Elghandour MMY, Salem AZM, Ganzalez-Ronquillo M et al (2013) Effects of exogenous enzymes on *in vitro* gas production kinetics and ruminal fermentation of four fibrous feeds. Anim Feed Sci Technol 179:46–53

El-Kady RI, Awadalla IM, Mohamed MI et al (2006) Effect of exogenous enzymes on the growth performance and digestibility of growing buffalo. Int J Agric Biotechnol 8:354–359

Elwakeel EA, Titgemeyer EC, Johnson BJ et al (2007) Fibrolytic enzymes to increase the nutritive value of dairy feedstuffs. J Dairy Sci 90:5226–5236

Engels FM, Schuurman JLL (1992) Relationship between structural development of cell walls and degradation of tissues in maize stems. J Sci Food Agric 59:45–51

Eun JS, Beauchemin KA (2007) Assessment of the efficacy of varying experimental exogenous fibrolytic enzymes using *in vitro* fermentation characteristics. Anim Feed Sci Technol 132:298–315

Faichney GJ (1986) Kinetics of particulate matter in the rumen. In: Milligan LP, Grovum WL, Dobson A (eds) Control of digestion and metabolism in ruminants, pp 173–195

FAO (2009) The state of food and agriculture. Livestock in the balance. FAO, Rome, p 168

Feng P, Hunt CW, Pritchard GT (1996) Effect of enzyme preparations on in situ and in vitro degradation and in vivo digestive characteristics of mature cool season grass forage in beef steers. J Anim Sci 74:1349–1357

Firkins JL, Allen MS, Oldick BS et al (1998) Modelling ruminal digestibility of carbohydrates and microbial protein flow to the duodenum. J Dairy Sci 81:3350–3369

Fisher DS, Burns JC, Moore JE (1995) The nutritive evaluation of forage. In: Miller BDA, Nelson CJ (eds) Forages: the science of grassland, vol I, 5th edn, pp 105–116

Forbes (2016) Urbanization will change the developing world. http://www.forbes.com/sites/danielrunde/2015/02/24/urbanization-development-opportunity/#797904d62779

Forsberg CW, Cheng KJ (1992) Molecular strategies to optimize forage and cereal digestion by ruminants. In: Bills DD, Kung SD (eds) Biotechnology and nutrition. Stoneham, UK, pp 107–147

France J, Dijkstra J (2005) Volatile fatty acid production. In: Dijkstra J, Forbes JM, France J (eds) Quantitative aspects of ruminant digestion and metabolism, 2nd edn. CABI, UK, pp 157–176

Gado HM, Salem AZM, Robinson PH et al (2009) Influence of exogenous enzymes on nutrient digestibility, extent of ruminal fermentation as well as milk production and composition in dairy cows. Anim Feed Sci Technol 154:36–46

Garnett T (2009) Livestock-related greenhouse gas emissions: impacts and options for policy makers. J Environ Sci Policy 12:491–503

Gibb DJ, McAllister TA (1999) The impact of feed intake and feeding behaviour of cattle on feedlot and feedbunk management. Agriculture and Agri-Food Canada, Lethbridge Research Centre, Lethbridge

Giraldo LA, Tejido ML, Ranilla MJ et al (2008a) Effects of exogenous fibrolytic enzymes on *in vitro* ruminal fermentation of substrates with different forage: concentrate ratios. Anim Feed Sci Technol 141:306–325

Giraldo LA, Tejido ML, Ranilla MJ et al (2008b) Influence of direct-fed fibrolytic enzymes on diet digestibility and ruminal activity in sheep fed a grass hay-based diet. J Anim Sci 86:1617–1623

Gordon GLR, Phillips MW (1992) Extracellular pectin lyase produced by *Neocallimastix* sp. LM1, a rumen anaerobic fungus. Lett Appl Microbiol 15:113–115

Grainger D, Beauchemin KA (2011) Can enteric methane emissions from ruminants be lowered without lowering their production? Anim Feed Sci Technol 166–167:308–320

Grant RJ (1997) Interactions among forages and non forage fibre sources. J Dairy Sci 80:1438–1446

Hatfield RD, Ralph J, Grabber JH (1999) Cell wall structural foundations: molecular basis for improving forage digestibilities. Crop Sci 39:27–37

Hindrichsen IK, Kreuzer M, Machmuller A et al (2003) Methane release and energy expenditure of dairy cows fed concentrates characterized by different carbohydrates. In: WM Souffrant, CC Metges (eds). Proceedings of the EAAP-symposium on energy and protein metabolism and nutrition 2003. pp. 413–416

Hristov A, McAllister TA, Cheng KJ (1998) Exogenous enzymes for ruminants: modes of action and potential applications. 17th Western nutrition conference, Edmonton, Alberta.

Hristov AN, Oh J, Firkins JL et al (2013) Mitigation of methane and nitrous oxide emissions from animal operations: II. A review of manure management mitigation options. J Anim Sci 91:5070–5094

Hull JL, Dobie JB, Morris JG (1974) Processing roughages for wintering steer calves. Calif Agric 28:10–11

Johnston JD (2000) Fibrozyme and *in vitro* NDF response: moving from theory to practical commercial reality. In: Lyons TP, Jacques K (eds) Biotechnology in the feed industry by 2000, vol 487. Alltech, Ontario. https://en.engormix.com/dairy-cattle/articles/fibrozyme-vitro-ndf-response-t33895.htm

Jung HG, Allen MS (1995) Characteristics of plant cell walls affecting intake and digestibility of forages by ruminants. J Anim Sci 73:2774–2790

Khan MJ, Abbas A, Ayaz M et al (2012) Factors affecting wool quality and quantity in sheep. Afr J Biotechnol 11:13761–13766

Kharas H (2010) The emerging middle class in developing countries. OECD Development Centre, Paris. Working paper no. 285

Kobayashi Y (2010) Abatement of methane production from ruminant: trends in the manipulation of rumen fermentation. Asian-Australian J Anim Sci 23:410–416

Kopecny J, Marounek M, Holub K (1987) Testing the suitability of the addition of *Trichoderma viride* cellulases to feed rations for ruminants. Zivoc Vyroba 32:587–592

Krause DO, Denman ST, Mackie RI et al (2003) Opportunities to improve fibre degradation in the rumen: microbiology, ecology, and genomics. FEMS Microbiol Rev 27:663–693

Krueger NA, Adesogan AT, Staples CR et al (2008a) The potential to increase digestibility of tropical grasses with a fungal, ferulic acid esterase enzyme preparation. Anim Feed Sci Technol 145:95–108

Krueger NA, Adesogan AT, Staples CR, Kreuger WK, Kim SC, Litell RC et al (2008b) Effects of method of applying fibrolytic enzymes or ammonia to Bermuda grass hay on feed intake, digestion and growth of beef steers. J Anim Sci 86:882–889

Kung JR, Treacher RJ, Nauman GA et al (2000) The effect of treating forages with fibrolytic enzymes on its nutritive value and lactation performance of dairy cows. J Dairy Sci 83:115–122

Leng RA (1991) Improving ruminant production and reducing methane emissions from ruminants by strategic supplementation. United States Environmental Protection Agency, Office of Air and Radiation, Washington, DC, p 105

Leng RA (2008) The potential of feeding nitrate to reduce enteric methane production in ruminants. A report to the Department of Climate Change Commonwealth Government of Australia, Canberra, ACT, Australia. www.penambulbooks.com

Loix C, Huybrechts M, Vangronsveld J, Gielen M, Keunen E, Cuypers A (2017) Reciprocal Interactions between Cadmium-Induced Cell Wall Responses and Oxidative Stress in Plants. Front Plant Sci 8:1867. https://doi.org/10.3389/fpls.2017.01867

Lopez-Soto MA, Plascencia A, Arellano GE et al (2000) Interaction of maceration and fibrolytic enzyme supplementation on the site and extent of digestion in rice straw in Holstein cows. West Sect Am Soc Anim Sci 51:458–462

Lovett D, Lovel LS, Stack L et al (2003) Effect of forage concentrate ratio and dietary coconut oil level on methane output and performance of finishing beef heifers. Livest Prod Sci 84:135–146

Marquardt RR, Brufau J (1997) Future of feed enzymes: orientation and perspectives. In: Morand-Fehr P (ed) Feed manufacturing in southern Europe: new challenges. CIHEAM, Zaragoza, pp 181–187

Martin C, Morgavi DP, Doreau M (2009) Methane mitigation in ruminants: from microbe to farm scale. Anim Consort 4:351–365

McAllister TA, Bae HD, Jones GA et al (1994) Microbial attachment and feed digestion in the rumen. J Anim Sci 7(2):3004–3018

McAllister TA, Okine EK, Mathison GW et al (1996) Dietary, environmental and microbiological aspects of methane production in ruminants. Can J Anim Sci 76:231–243

McAllister TA, Hristov AN, Beauchemin KA et al (2001) Enzymes in ruminant diets. In: Bedford MR, Partridge GG (eds) Enzymes in farm animal nutrition. CAB International, Wiltshire

McDonald P, Edwards R, Greenhalgh JFD et al (2011) Animal nutrition, 7th edn. Harlow, Pearson Education, Prentice Hall, England

McGeough EJ, O'ki ELYP, Hart KJ et al (2010) Methane emissions, feed intake, performance, digestibility, and rumen fermentation of finishing beef cattle offered whole-crop wheat silages differing in grain content. J Anim Sci 88:2703–2716

McGinn SM, Beauchemin KA, Coates T et al (2004) Methane emissions from beef cattle: effects of monensin, sunflower oil, enzymes. Yeast and fumaric acid. J Anim Sci 82:3346–3356

McManus WR, Bigham ML (1973) Studies on forage cell walls: prediction of feed intake. J Agric Sci 80:283–296

McNeill M, Darvill AG, Fry SC, Albertsheim P (1984) Structure and function of the primary cell wall of plants. Annu Rev Biochem 53:625–663

Meissner HH (1997) Recent research on forage utilization by ruminant livestock in South Africa. Anim Feed Sci Technol 69:103–119

Mittler R, Blumwald E (2010) Genetic engineering for modern agriculture: challenges and perspectives. Ann Rev Plant Biol 61:443–462

Mohamed DEA, Borhami BE, El-Shazly KA et al (2013) Effect of dietary supplementation with fibrolytic enzymes on the productive performance of early lactating dairy cows. J Agric Sci 5:146–155

Monteny GJ, Chadwick D (2006) Greenhouse gas abatement strategies for animal husbandry. Agric Ecosyst Environ 112:163–170

Morgavi DP, Beauchemin KA, Nsereko VL et al (2000a) Synergy between ruminal fibrolytic enzymes and enzymes from *Trichoderma longibrachiatum*. J Dairy Sci 83:1310–1321

Morgavi DP, Newbold CJ, Beever DJ et al (2000b) Stability and stabilization of potential feed additive enzymes in rumen fluid. Enz Microbiol Technol 26:171–177

Morgavi DP, Forano E, Martin C et al (2010) Microbial ecosystems and methanogenesis in ruminants. Anim Sci 4:1024–1036

Murillo M, Alvarez EG, Cruz J et al (2000) Interaction of forage level and fibrolytic enzymes on digestive function in cattle. West Sect Am Soc Anim Sci 51:324–326

Murray RM, Bryant AM, Leng RA (1976) Rates of production of methane in the rumen and large intestine of sheep. Br J Nutr 36:1–14

Ndlovu LR, Buchanan-Smith JG (1985) Utilization of poor quality roughages by sheep: effects of alfalfa supplementation on ruminal parameters, fibre digestion and rate of passage from the rumen. Can J Anim Sci 65:693–703

Ochoa-Villarreal M, Aispuro-Hernández E, Vargas-Arispuro I et al (2012) Plant cell wall polymers: function, structure and biological activity of their derivatives. In: Gomes ADS (ed) Polymerization, vol 4, pp 63–86

Perry TW, Purkhiser ED, Beeson WM (1966) Effects of supplemental enzymes on nitrogen balance, digestibility of energy and nutrients and on growth and feed efficiency of cattle. J Anim Sci 25:760–764

Pinos-Rodriguez JM, Gonzalez SS, Mendoza GD et al (2002) Effect of exogenous fibrolytic enzyme on ruminal fermentation and digestibility of alfalfa and rye-grass hay fed to lambs. J Anim Sci 80:3016–3020
Plaizier JC, Krause DO, Gozho GN et al (2009) Subacute ruminal acidosis in dairy cows: the physiological causes, incidence and consequences. J Vet Sci 176:21–31
Rust JW, Jacobsen NL, McGilliard AD et al (1965) Hotchkiss DK. Supplementation of dairy calf diets with enzymes. 1. Effect on nutrient utilization and on the composition of rumen fluid. J Anim Sci 24:156–160
Santoso B, Kume S, Nonaka K et al (2003) Methane emission, nutrient digestibility, energy metabolism and blood metabolites in dairy cows fed silages with and without galacto-oligosaccharides supplementation. Asian-Australian J Anim Sci 16:534–540
Schingoethe DJ, Stegeman GA, Treacher RJ (1999) Response of lactating dairy cows to a cellulase and xylanase enzyme mixture applied to forages at the time of feeding. J Dairy Sci 82:996–1003
Selinger LB, Forsberg CW, Cheng KJ (1996) The rumen: a unique source of enzymes for enhancing livestock production. Anaerobe 2:263–284
Sheppy C (2001) The current feed enzyme market and likely trends. In: Bedford MR, Partridge GG (eds) Enzymes in farm animal nutrition, pp 1–9
Singh B, Sahoo A, Sharma R et al (2005) Effect of polyethylene glycol on gas production parameters and nitrogen disappearance of some tree forages. Anim Feed Sci Technol 123–124:351–364
Smith AH, Zoetendal EG, Mackie M (2005) Bacterial mechanisms to overcome inhibitory effects of dietary tannins. Microb Ecol 50:197–205
Sorek N, Yeats TH, Szemenyei H et al (2014) The implications of lignocellulosic biomass chemical composition for the production of advanced biofuels. Biol Sci 64:192–201
Steinfeld H, Gerber P, Castel V et al (2006) Livestock's long shadow, vol 2006. FAO, Rome
Sutton JD, Phipps RH, Beever DE et al (2003) Effect of method of application of a fibrolytic enzyme product on digestive processes and milk production in Holstein-Friesian cows. J Dairy Sci 86:546–556
Theurer B, Woods W, Burroughs W (1963) Influence of enzyme supplements in lamb fattening rations. J Anim Sci 22:150–154
Travis AJ, Murison SD, Perry P et al (1997) Measurement of cell wall volume using confocal microscopy and its application to studies of forage degradation. Ann Bot 80:1–11
Tyrrell HF, Moe PW (1975) Effect of intake on digestive efficiency. J Dairy Sci 58:1151–1163
Ushida K, Newbold CJ, Jouany JP (1997) Interspecies hydrogen transfer between the rumen ciliate *Polyplastronmulti vesiculatum* and *Methanosarcina barkeri*. J Gen Appl Microbiol 43:129–131
Van Soest PJ (1994) Nutritional ecology of the ruminant, 2nd edn. Cornell University Press, Ithaca
Van Soest PJ, Mertens DR, Deinum B (1978) Preharvest factors influencing quality of conserved forage. J Anim Sci 47:712–720
Varga GA, Kolver ES (1997) Microbial and animal limitations to fibre digestion and utilization. Am Soc Nutr Sci 127:819S–823S
Wallace RJ, Wallace SJ, McKain N et al (2001) Influence of supplementary fibrolytic enzymes on the fermentation of corn and grass silages by mixed ruminal microorganisms *in vitro*. J Anim Sci 79:1905–1916
Walters RJK (1971) Variation in the relationship between in vitro digestibility and voluntary dry matter intake of different grass varieties. J Agric Sci 76:243–252
Wanapat M, Chenost M, Munoz F et al (1996) Methods for improving the nutritive value of fibrous feed: treatment and supplementation. Ann Zootech INRA/EDP Sci 45:89–103
Wang D, Zhang H, Wu F et al (2013) Modification of pectin and hemicellulose polysaccharides in relation to aril breakdown of harvested Longan fruit. Int J Mol Sci 14:23356–23368
Ware RA, Alvarez EG, Machado M et al (2002) Influence of pelletizing on the feeding value of rice straw in growing–finishing diets. West Sect Am Soc Anim Sci 53:637–641
Weimer PJ (1996) Why don't ruminal bacteria digest cellulose faster? J Dairy Sci 79:1496–1502

Weimer PJ, Russell JB, Muck RE (2009) Lessons from the cow: what the ruminant animal can teach us about consolidated bioprocessing of cellulosic biomass. Bioresour Technol 100:5323–5331

Wenk C (2000) Recent advances in animal feed additives such as metabolic modifiers, antimicrobial agents, probiotics, enzymes and highly available minerals. Asian Australasian J Anim Sci 13:86–95

Wilkins RJ (2000) Forages and their role in animal systems. In: Givens DL, Owen E, Axford REF, Omed HM (eds) Forage evaluation in ruminant nutrition. CABI, Wallingford, pp 1–12

Wilson JR (1993) Organisation of forage plant tissues. In: Jung HG, Buxton DR, Hatfield RD, Ralph J (eds) Forage cell wall structure and digestibility, Am Soc Agr, vol 1993, pp 1–32

Wilson JR, Mertens DR (1995) Crop quality and utilization. Cell wall accessibility and cell structure limitations to microbial digestion of forage. Crop Sci 35:251–259

Wood TM, Wilson CA, McCrae SI et al (1986) Joblin KN. A highly active extracellular cellulase from the anaerobic rumen fungus *Neocallimastix frontalis*. FEMS Microbiol Lett 34:37–40

Worldwatch (2016) Global meat production and consumption continue to rise. http://www.worldwatch.org/global-meat-production-and consumption-continue-rise-1

Yang WZ, Beauchemin KA (2001) Effects of grain processing, forage to concentrate ratio, and forage particle size on rumen pH and digestion by dairy cows. J Dairy Sci 84:2203–2216

Yang WZ, Beauchemin KA, Rode LM (1999) Effects of an enzyme feed additive on the extent of digestion and milk production of lactating dairy cows. J Dairy Sci 82:391–403

Yang WZ, Beauchemin KA, Rode LM (2000) A comparison of methods of adding fibrolytic enzymes to lactating cow diets. J Dairy Sci 83:2512–2520

Zhong W, Zhang Z, Luo Y et al (2011) Effect of biological pretreatments in enhancing corn straw biogas production. Bioresour Technol 102:11177–11182

Inclusion of Exogenous Fibrolytic Enzymes in the Diets of Dairy Cows and Ewes: Effect on Milk Yield and Milk Composition

Lizbeth E. Robles Jimenez, Babak Darabighane, Sergio Radic-Schilling, Carlos Palacios, Alfonso J. Chay Canul, Ricardo A. Garcia-Herrera, and Manuel Gonzalez-Ronquillo

Abstract Supplementation of exogenous enzymes in ruminant diets is an alternative solution to increase dietary energy and fiber digestibility and thus decrease production costs. A systematic review was conducted, resulting from 21 studies with 76 experimental records, divided by animal species (dairy cows or dairy sheep). A descriptive statistical analysis of milk production and milk composition was carried out. It was found that xylanases or the combination of xylanases with cellulases produced more milk and improved the percentage of fat, lactose, protein, and total solids. Future studies should focus on the composition of the diet, the origin of the enzyme, and the amount and number of enzymes added.

L. E. Robles Jimenez · A. J. Chay Canul · R. A. Garcia-Herrera
Universidad Juárez Autónoma de Tabasco, División Académica de Ciencias Agropecuarias, Carretera Villahermosa-Teapa, Villahermosa, Tabasco, Mexico

B. Darabighane
Ministry of Agriculture, Tehran, Iran

S. Radic-Schilling
University of Magallanes, Faculty of Sciences , Punta Arenas, Chile
e-mail: sergio.radic@umag.cl

C. Palacios
University of Salamanca, Faculty of Environmental and Agricultural Sciences Animal Production Group. Department of Construction and Agronomy, Salamanca, Spain
e-mail: carlospalacios@usal.es

M. Gonzalez-Ronquillo (✉)
Universidad Autónoma del Estado de México, Facultad de Medicina Veterinaria y Zootecnia, Department of Animal Production, Toluca, Estado de Mexico, Mexico
e-mail: mrg@uaemex.mx

A. Z. M. Salem et al. (eds.), *Exogenous Enzymes as Feed Additives in Ruminants*, https://doi.org/10.1007/978-3-031-27993-5_6

1 Introduction

The animal's digestive system is not fully efficient in digesting the total diet provided, because feed ingredients contain nondigestible factors that hinder the digestive process, and/or the animal lacks the enzymes necessary to degrade certain feed components (Cowieson et al. 2006; Selle and Ravindran 2007). All animals use endogenous enzymes to digest feed, which are produced by the animal's system or by microorganisms that naturally inhabit the intestine; however, the animal's digestive system is not fully efficient (Selle and Ravindran 2007). To improve the nutritional value of feed and remove undesirable factors (e.g., fiber, phytate), exogenous enzymes are used in animal feed (Ojha et al. 2019).

Exogenous enzymes have been used to improve nutrient utilization in various species but are mostly used in the poultry and swine industries; however, in ruminants, they have been reported to have been implemented since 1960 (Burroughs et al. 1960). Nowadays, the current renewed interest in the use of enzymes for ruminants has been mainly due to the costs of fermentation/generation, as well as new technologies to produce purified enzyme preparations at a lower cost (Zilio et al. 2019).

There are generally four categories of enzymes marketable by the animal feed industry for monogastric animals: (1) enzymes targeting phytates, (2) enzymes targeting viscous cereals (i.e., *Triticum*), (3) enzymes targeting nonviscous cereals (i.e., corn), and (4) enzymes targeting noncereals/legumes (i.e., flax meal, sunflower) (Ojha et al. 2019).

Enzymes have different modes of action, and for some, the mode of action remains unknown (Bedford and Partridge 2011; Ravindran 2013).

Exogenous feed enzymes help to:

(a) Degrade exact bonds in feed ingredients/components that are not normally hydrolyzed by endogenous enzymes.
(b) Degrade deleterious factors present in the feed that are responsible for reducing digestion and increasing intestinal viscosity, thus achieving better utilization of nutrients.
(c) They can interfere with the integrity of the endosperm and cause the discharge of nutrients that are bound or trapped by the plant cell wall. They modify the digestion process to favor digestion sites.
(d) Reduce internal secretions and protein loss from the digestive tract, thereby reducing the maintenance needs of animals (Cowieson et al. 2009).
(e) Reduce intestinal load and modify intestinal morphology (Wu et al. 2004).
(f) They change the microbial populations of the gastrointestinal (GI) tract directly by manipulating the amount and appearance of substrate present in the GI tract due to an increase in beneficial bacteria or decrease in detrimental microbes in the gut (Apajalahti et al. 2004; Bedford and Cowieson 2012).

The exogenous enzymes mainly used in the animal feed industry as feed additives are cellulase, β-glucanases, xylanases, and associated enzymes phytases, proteases, lipases, and galactosidases (Table 1).

Table 1 Type of exogenous enzymes used in animal feed industry and their substrates

Enzyme	Substrate	Feedstuff
α-Galactosidases	Oligosaccharides	Linseed meal, leguminous grain
β-Glucanases	β-Glucans	Barley, oats, rye
Amylases	Starch	cereal grains, leguminous grain
Lipases	Lipids	Fats and oils in feed sources
Mannanases, cellulases, Hemicellulases, pectinases	Cell wall matrix (crude fiber)	Plant-origin fibrous ingredients
Phytases	Phytate	Plant-based feed sources
Proteases	Proteins	Plant-derived protein ingredients
Xylanases	Arabinoxylans	Wheat, triticale, fibrous plant complexes

Adapted from Ojha et al. (2019)

1.1 Uses of Enzymes in Ruminants

Forages are the main source of feed for ruminants (Kholif et al. 2014). However, fibrous feeds can be characterized as having low nutritional value, due to the high fiber and silica content in their walls, which reduces the access of ruminal enzymes to the plant cell walls, thus reducing their digestibility (Kholif et al. 2014). Despite improvements achieved through forage selection and hybridization programs, cell wall digestibility is still a limiting factor in terms of utilization of the energy intake available to ruminants (Beauchemin et al. 2003). The most used exogenous enzymes in ruminant diets can be classified into three categories: fibrolytic, amylolytic, and proteolytic (Sujani and Seresinhe 2015).

The use of exogenous fibrolytic enzymes has been identified as one of the alternatives to increase forage utilization and improve productive efficiency (González-García et al. 2010).

The supplementation of fibrolytic enzymes in ruminant diets is attractive, since fiber fractions can be digested much more efficiently: the digestibility of fiber fractions peaks at 65–70% (Sujani and Seresinhe 2015).

Exogenous fibrolytic enzymes for ruminants generally contain fermentation products of fungi (usually *Trichoderma longibrachiatum*, *Aspergillus niger*, A. *oryzae*) or bacteria (*Bacillus* spp.) (Fig. 1) (The Association of American Feed Control Officials 2002).

Within the rumen, exogenous fibrolytic enzymes hydrolyze certain components of the forage cell wall and produce substrates that enhance the development of selected populations of microorganisms, even in low forage diets (Bedford and Cowieson 2012).

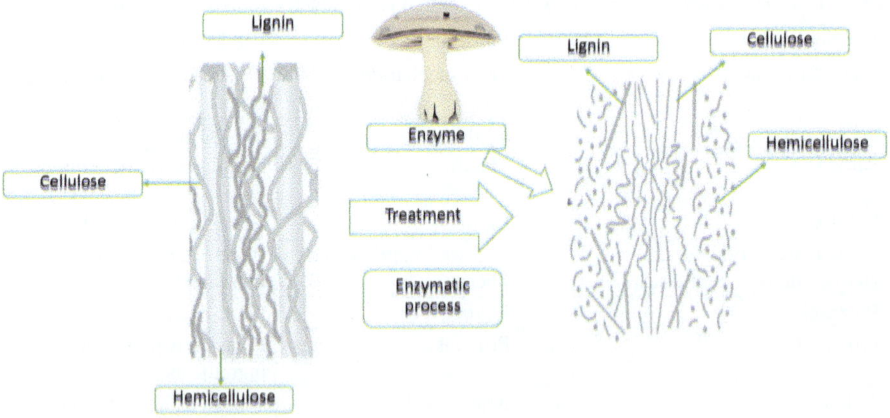

Fig. 1 Use of fibrolytic enzymes extracted from fungi for animal feed

Exogenous fibrolytic enzymes used in ruminates can be classified based on their specific activity as: cellulases (which hydrolyze cell wall fiber into glucose, cellobiose or cello-oligosaccharides) through the action of three enzymes: endoglucanase, exoglucanase, and X-glucosidates (Zhang and Lynd 2004). Xylanases are named so, because they hydrolyze the 1,4 beta-D-xylosidic bonds of xylan, which are constituents of hemicellulose (a structural component of plant cell walls) (Kamble and Jadhav 2012). Its main enzymes are endo-X-1,4-xylanase (1,4-XD xylan hydrolyase), which attacks the main chain of xylan and XD-xylosidase (1,4-X-xylan hydrolyase), which hydrolyzes xylooligosaccharides into D-xylose, in addition to a variety of debranching enzymes, i.e., "L-arabino-furanosidases", L-arabinoglucuronidases, and acetyl esterases (Kamble and Jadhav 2012). The activity of fiber-degrading enzymes is determined by measuring the rate of release of reduced sugars, divided per unit time per unit enzyme (Beauchemin et al. 2003). Xylanases have a specific function against β-1,4-linkages within the xylan backbone and are generally considered endoxylanases (Bhat and Hazlewood 2009).

Alpha-amylases are radical innovations for dairy cattle nutrition and can be used to maximize starch utilization and fiber utilization. These enzymes act on the rumen by enhancing starch hydrolysis and catalyzing the reaction that converts starch to oligosaccharides, without affecting pH (Fig. 2) (nutriNews 2020).

Due to the benefits observed with these enzymes, several studies have added exogenous enzymes in ruminant diets. Therefore, the objective of this study was to perform a systematic review to evaluate the effect of the addition of exogenous enzymes (xylanases and cellulases) in ruminant feed on milk production and composition.

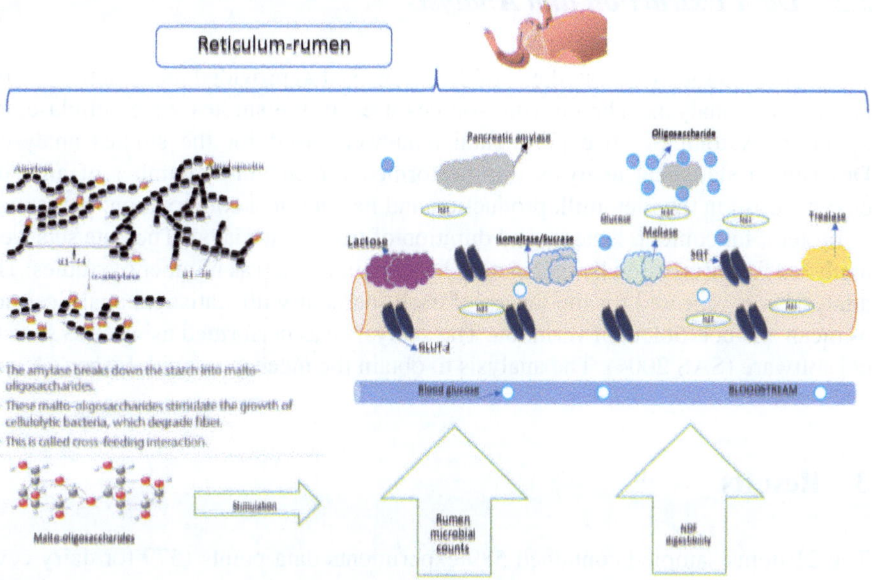

Fig. 2 Mechanism of action of alpha-amylase in the rumen

2 Materials and Methods

2.1 *Search Strategy and Selection Criteria*

Our search for information focused on studies reporting on the use of exogenous enzymes in ruminant diets. A database was created from studies specifying the use of exogenous enzymes in diets (dairy cows and ewes) and the articles used, covering the years 1999–2021. Publications were obtained from databases such as Worldwide Science, ScienceDirect, Scopus, Springer Link, Wiley Online Library, Dialnet, SciELO, Science Research, PubMEd, Redalyc and Google Academic. Information retrieval to find relevant publications was based on a specific topic string, such as exogenous enzymes (xylanases and cellulases) used in ruminants.

The search string with the topic was supported by Boolean operators ("and," "or"), which served to specify the required information. All search terms within a string were checked for a "title, abstract, and keyword," exogenous enzymes, ruminants, milk production, "enzymes and exogenous and ruminants," "enzymes and milk production," "enzymes and ruminants or dairy cattle or dairy ewes." In addition, the following variables were recorded in the database: number of animals in the study, basal diet, enzyme used and its source, commercial name of the enzyme, route of administration, enzyme dose (g/kg Live weight, $LW^{0.75}$), milk production (kg/kg $LW^{0.75}$), and its milk composition: protein, fat, lactose and total solids content (g/100 g), and duration of treatment (days).

2.2 Data Extraction and Analysis

We obtained a total of 21 studies (Table 2, Fig. 3) that provided the data to form the basis of the analysis. The enzyme sources used in the studies were cellulase and xylanase. A total of 76 experimental data were used for the studies analyzed. Descriptive statistical analysis was performed for enzymes, number of animals, enzyme dose in the diet, milk production and its chemical composition: percentage of protein, fat content, lactose, and duration of treatment (days). The data sets were analyzed for bifurcation by calculating basic indices such as number of studies. The analysis was repeated for the length of each segment with statistical analyses such as mean and coefficient of variation. The analysis was performed using SAS statistical software (SAS 2004). The analysis to obtain the means was with Fisher's F test.

3 Results

The 21 items sampled contained 589 experiments data points (579 for dairy cows and 10 for dairy ewes). The average metabolic body weight of the cows was 125 kg $LW^{0.75}$, the treatments supplying xylanase dose (6.12 g/kg) in dairy cows obtained the highest amount of milk (39.67 kg/d), lactose (4.81%), protein (3.20%), and total solids (12.70%); however, the time of cellulase supply was longer (80 d). The treatments that supplied the dose of cellulose-xylanase (0.43 g/kg) in ewes obtained the highest quantity of milk (2.40 kg/d) and the period of supply was shorter (28 d); however, the highest quantity of fat (7.33%), protein (6.11%), and total solids (18.33%) was obtained with the use of cellulose (Table 3).

4 Discussion

Supplementation with fibrolytic enzymes (xylanase, cellulase, and xylanase-cellulase) increased milk yield and milk solids. These results coincide with those reported in the meta-analyses of Kholif and Aziz (2014), Mohamed et al. (2013), and Tirado-González et al. (2018), who found that the use of a preparation with mainly xylanases activity increased milk production by 1.5 kg/d (3.8%); however, they also mention that the amount of forage has to be increased. This increase may be possible due to the fact that fibrolytic enzymes increase the amount of digestible energy of diets that are high in fiber (Meale et al. 2014).

When we look at the results obtained by using enzyme combinations in both, cows and sheep, we can see that they can perform better than a pure enzyme (Yu et al. 2005). This may be because the right mix of xylanases and cellulases improves glucose release (Eun and Beauchemin 2007). However, in many studies, they do not contain the right mix of enzymes, so the results are not consistent. Excess of

Table 2 Type of enzymes and diet used in dairy cows and dairy ewes

Study	Reference	Species	Basal feed	Enzyme
1	Titi and Lubbadeh (2004)	Dairy ewe	Alfalfa hay, concentrate	Cellulase
2	Flores et al. (2008)	Dairy ewe	Maize-whole plant, alfalfa hay, concentrate	Cellulase, xylanase
3	Bowman et al. (2002)	Dairy cattle	Barley silage, alfalfa silage, concentrate mixture	Cellulase, xylanase
4	Dhiman et al. (2002)	Dairy cattle	Alfalfa Hay, corn silage, concentrate	Cellulase
5	Gado et al. (2009)	Dairy cattle	Corn silage, concentrate	Cellulase xylanase
6	Knowlton et al. (2002)	Dairy cattle	Alfalfa silage, corn silage, concentrate	Cellulase
7	Kung et al. (2000)	Dairy cattle	Corn silage, alfalfa hay, pelleted concentrate	Cellulase, xylanase
8	Kung et al. (2002)	Dairy cattle	Corn silage, alfalfa hay, pelleted concentrate	Cellulase, xylanase
9	Lewis et al. (1999)	Dairy cattle	Alfalfa silage, alfalfa hay, concentrate	Cellulase, xylanase
10	Pinos-Rodríguez et al. (2005)	Dairy cattle	Corn silage, alfalfa hay, concentrate	Xylanase
11	Reddish and Kung (2007)	Dairy cattle	Corn silage, alfalfa silage, alfalfa hay, concentrate	Cellulase, xylanase
12	Rode et al. (1999)	Dairy cattle	Corn silage, alfalfa hay, concentrate	Cellulase, xylanase
13	Sutton et al. (2003)	Dairy cattle	Corn silage, alfalfa hay, alfalfa silage, concentrate	Xylanase
14	Yang et al. (1999)	Dairy cattle	Barley silage, alfalfa hay, concentrate	Cellulase, xylanase
15	Yang et al. (2000)	Dairy cattle	Corn silage, alfalfa hay, concentrate	Cellulase, xylanase
16	Zheng et al. (2000)	Dairy cattle	Corn silage, alfalfa hay, concentrate	Xylanase
17	Beauchemin et al. (1999)	Dairy cattle	Barley silage, alfalfa haylage, concentrate	Xylanase
18	Elwakeel et al. (2007)	Dairy cattle	Corn silage, alfalfa hay, concentrate	Xylanase
19	Arriola et al. (2011)	Dairy cattle	Corn silage, alfalfa hay, concentrate	Cellulase, xylanase
20	Knowlton et al. (2007)	Dairy cattle	Corn silage, legume-grass silage, concentrate	Cellulase
21	Bernard et al. (2010)	Dairy cattle	Corn silage, alfalfa hay, concentrate	Cellulase

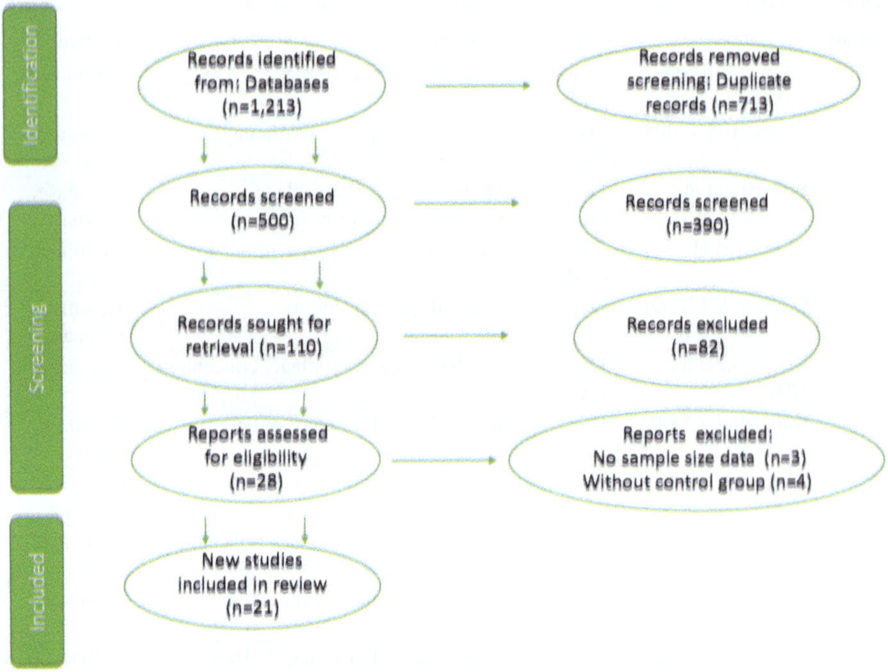

Fig. 3 PRISMA study flow diagram of the systematic review from the initial search and selection to the final selection of the publications to be included in the study

cellulase may limit carbohydrate hydrolysates and reduce microbial adhesion (Morgavi et al. 2001; Wang et al. 2004); excess of xylanase activities may also negatively affect the population of rumen microorganisms (Eun and Beauchemin 2008).

Furthermore, optimal doses of fibrolytic enzymes would vary according to diet composition (Fortes et al. 2010). For example, legume-based diets may contain more enzymes than grass-based diets (Yang et al. 2011). These differences in results can also be influenced by the following factors: inclusion dose, enzyme composition, type of diet, previous adaptation, time-period used, and methods of addition of enzyme complexes to diets. In addition, researchers have shown that the increase in feed digestibility is not solely due to the simple result of supplementation of fibrolytic enzymes and their enzyme activity, as their contribution to the overall enzyme activity of the rumen is relatively small (McAllister et al. 2009).

Although in the present analysis we found some positive effects with the use of enzymes in dairy sheep in some studies, for example, improved weight gain, increase in propionic acid, and volatile fatty acids ration were found due to enzyme supplementation (McAllister et al. 2000; Miller et al. 2008; Gado et al. 2011), others, however, reported negative and inconsistent results with the use of enzymes in dairy ewes (Tirado-González et al. 2018).

Table 3 Effects of adding primarily cellulases, cellulases:xylanases mixture, and xylanases as an exogenous fibrolytic enzymes to different diets on animal performance

			Enzymatic activity		
Variable	*N*	Cel	Xyl	Cel:Xyl	Control
Dairy cows	579				
$LW^{0.75}$		125 ± 6.0	124 ± 6.2	124 ± 7.0	124 ± 6.2
DMI kg/d		23.90 ± 1.0	25.92 ± 1.5	23.23 ± 1.2	24.04 ± 1.0
Enzyme dose g/kg		2.83 ± 0.1	6.12 ± 0.3	4.87 ± 0.2	0
Time of use, d		80 ± 4	61 ± 3	56 ± 3	61 ± 3
Milk per day, kg		37.95 ± 2.0	39.67 ± 2.1	33.77 ± 1.6	33.76 ± 1.5
Fat %		2.97 ± 0.2	3.60 ± 0.1	3.70 ± 0.1	4.5 ± 0.2
Lactose %		1.97 ± 0.05	4.81 ± 0.2	4.67 ± 0.1	3.93 ± 0.2
Protein %		1.90 ± 0.08	3.20 ± 0.1	3.08 ± 0.1	2.93 ± 0.1
Total solids %			12.70 ± 0.6		10.15 ± 0.1
Dairy ewes	10				
$MBW^{0.75}$		21.56 ± 1.1		25.33 ± 1.3	24 ± 1.2
DMI kg/d		2.28 ± 0.11		2.99 ± 0.14	2.81 ± 0.13
Enzyme dose g/kg		0.15 ± 0.001		0.47 ± 0.02	0
Time of use, d		120 ± 4		28 ± 1	51 ± 2
Milk per day, kg		2.42 ± 0.02		2.40 ± 0.1	1.91 ± 0.1
Fat %		7.33 ± 0.4		6.09 ± 0.3	6.05 ± 0.3
Protein %		6.11 ± 0.3		5.30 ± 0.2	5.26 ± 0.2
Total solids %		18.33 ± 0.2		17.41 ± 0.7	17.03 ± 0.8

N number of observations, *Cel* primarily cellulases, *Xyl* primarily xylanases, *Cel:Xyl* mixture of cellulases:xylanases, $LW^{0.75}$ Metabolical Live Weight, *DMI* dry matter intake

5 Conclusions

Supplementation with fibrolytic enzymes improved milk production and milk composition (fat, protein, total solids, and lactose) in both cows and sheep; however, the results have great variability, and more studies are required to understand the interaction between diet composition, type of enzyme, and amount fed, as all of these variables interfere with the mode of action. Although most research using exogenous enzyme supplements in ruminant diets has produced positive results compared to control diets, there are still many inconsistencies in the results recorded and dosage that was used. This deserves more attention in the future.

Acknowledgments Miss. LERJ was granted with a CONACyT, Estancias posdoctorales 2021.

References

Apajalahti J, Kettunen A, Graham H (2004) Characteristics of the gastrointestinal microbial communities, with special reference to chickens. Worlds Poult Sci J 60:223–232

Arriola KG, Kim SC, Staples CR, Adesogan AT (2011) Effect of fibrolytic enzyme application to low- and high-concentrate diets on the performance of lactating dairy cattle. J Dairy Sci 94:832–841

Beauchemin KA, Yang WZ, Rode LM (1999) Effects of grain source and enzyme additive on site and extent of nutrient digestion in dairy cows. J Dairy Sci 82:378–390

Beauchemin K, Colombatto D, Morgavi DP, Yang WZ (2003) Use of exogenous fibrolytic enzymes to improve feed utilization by ruminants. J Anim Sci 81(14_suppl_2):E37–E47. https://doi.org/10.2527/2003.8114_suppl_2E37x

Bedford MR, Cowieson AJ (2012) Exogenous enzymes and their effects on intestinal microbiology. Anim Feed Sci Technol 173:76–85

Bedford MR, Partridge GG (2011) Enzymes in farm animal nutrition, 2nd edn. CAB International, Wallingford

Bernard JK, Castro JJ, Mullis NA, Adesogan AT, West JW, Morantes G (2010) Effect of feeding alfalfa hay or Tifton 85 bermudagrass haylage with or without a cellulase enzyme on performance of Holstein cows. J Dairy Sci 93:5280–5285

Bhat MK, Hazlewood GP (2009) Enzymology and other characteristics of cellulases and xylanases. Enzymes Farm Anim Nutr 11–60. https://doi.org/10.1079/9780851993935.0011

Bowman GR, Beauchemin KA, Shelford JA (2002) The proportion of the diet to which fibrolytic enzymes are added affects nutrient digestion by lactating dairy cows. J Dairy Sci 85:3420–3429

Burroughs W, Woods W, Ewing SA, Greig J, Theurer B (1960) Enzyme additions to fattening cattle rations. J Anim Sci 19:458–464. https://doi.org/10.2527/jas1960.192458x

Cowieson AJ, Acamovic T, Bedford MR (2006) Supplementation of corn–soy-based diets with an *Escherichia coli*-derived phytase: effects on broiler chick performance and the digestibility of amino acids and metabolizability of minerals and energy. Poult Sci 85:1389–1397

Cowieson AJ, Bedford MR, Selle PH, Ravindran V (2009) Phytate and microbial phytase: implications for endogenous nitrogen losses and nutrient availability. Worlds Poult Sci J 65:401–418

Dhiman TR, Zaman MS, Gimenez RR, Walters JL, Treacher R (2002) Performance of dairy cows fed forage treated with fibrolytic enzymes prior to feeding. Anim Feed Sci Technol 101:115–125

Elwakeel EA, Titgemeyer EC, Johnson BJ, Armendariz CK, Shirley JE (2007) Fibrolytic enzymes to increase the nutritive value of dairy feedstuffs. J Dairy Sci 90:5226–5236

Eun JS, Beauchemin KA (2007) Use in vitro degradation of alfalfa hay and corn silage using feed enzymes. J Dairy Sci 90:2839–2851. https://doi.org/10.3168/jds.2006-820

Eun JS, Beauchemin KA (2008) Assessment of the potential of feed enzyme additives to enhance utilization of corn silage fibre by ruminants. Can J Anim Sci 88:97–106. https://doi.org/10.4141/CJAS07042

Flores C, Caja G, Casals R, Albanell E, Such X (2008) Performance of dairy ewes fed diets with a fibrolytic enzyme product included in the concentrate during the suckling period. Animal 2(6):962–968

Fortes GLM, Alves OR, Silva BEP (2010) Cellulases, xylanases, B-glucosidases and ferulic acid esterases produced by Trichoderma and Aspergillus act synergistically in the hydrolysis of sugarcane bagasse. Biochem Eng J 51:72–78. https://doi.org/10.1016/j.bej.2010.05.003

Gado HM, Salem AZM, Robinson PH, Hassan M (2009) Influence of exogenous enzymes on nutrient digestibility, extent of ruminal fermentation as well as milk production and composition in dairy cows. Anim Feed Sci Technol 154:36–46

Gado HM, Salem AZM, Odongo NE, Borhami BE (2011) Influence of exogenous enzymes ensiled with orange pulp on digestion and growth performance in lambs. Anim Feed Sci Technol 165:131–136. https://doi.org/10.1016/j.anifeedsci.2011.02.016

González-García E, Albanell E, Caja G, Casals R (2010) In vitro fermentative characteristics of ruminant diets supplemented with fibrolytic enzymes and ranges of optimal endo-β-1,4-glucanase activity. J Anim Physiol Anim Nutr 94(2):250–263. https://doi.org/10.1111/j.1439-0396.2008.00903.x

Kamble DR, Jadhav AR (2012) Isolation, identification and screening of potential cellulase-free xylanase producing fungi and its production. Afr J Biotechnol 11(77):4624–4630. https://doi.org/10.5897/AJB11.2356

Kholif AM, Aziz HA (2014) Influence of feeding cellulolytic enzymes on performance, digestibility, and ruminal fermentation in goats. Anim Nutr Feed Technol 14(1):121–136

Kholif AE, Khattab HM, El-Shewy AA, Salem AZM, Kholif AM, El-Sayed GHM, Mariezcurrena MD (2014) Nutrient digestibility, ruminal fermentation activities, serum parameters and milk production and composition of lactating goats fed diets containing rice straw treated with Pleurotus ostreatus. Asian Australasian J Anim Sci 27(3):357–364. https://doi.org/10.5713/AJAS.2013.13405

Knowlton KF, McKinney JM, Cobb C (2002) Effect of a direct-fed fibrolytic enzyme formulation on nutrient intake, partitioning, and excretion in early and late lactation Holstein cows. J Dairy Sci 85:3328–3335

Knowlton KF, Taylor MS, Hill SR, Cobb C, Wilson KF (2007) Manure nutrient excretion by lactating cows fed exogenous phytase and cellulase. J Dairy Sci 90:4356–4360

Kung L, JrRJ T, Nauman GA, Smagala AM, Endres KM, Cohen MA (2000) The effect of treating forages with fibrolytic enzymes on its nutritive value and lactation performance of dairy cows. J Dairy Sci 83:115–122

Kung L, JrMA C, Rode LM, Treacher RJ (2002) The effect of fibrolytic enzymes sprayed onto forages and fed in a total mixed ratio to lactating dairy cows. J Dairy Sci 85:2396–2402

Lewis GE, Sanchez WK, Hunt CW, Guy MA, Pritchard GT, Swanson BI, Treacher RJ (1999) Effect of direct-fed fibrolytic enzymes on the lactational performance of dairy cows. J Dairy Sci 82:611–617

McAllister TA, Stanford K, Bae HD, Treacher RJ, Hristov AN, Baah J, Shelford JA, Cheng KJ (2000) Effect of surfactant and exogenous enzymes on digestibility of feed and on growth performance and carcass traits of lambs. Can J Anim Sci 80:35–44. https://doi.org/10.4141/A99-053

McAllister TA, Hristov AN, Beauchemin KA, Rode L, Cheng KJ (2009) Enzymes in ruminant diets. Enzyme Farm Anim Nutr 273–298. https://doi.org/10.1079/9780851993935.0273

Meale SJ, Beauchemin KA, Hristov AN, Chaves AV, McAllister TA (2014) Invited review: opportunities and challenges in using exogenous enzymes to improve ruminant production. J Anim Sci 92(2):427–442. https://doi.org/10.2527/jas.2013-6869

Miller DR, Granzin BC, Elliot R, Norton BW (2008) Effect of an exogenous enzyme Roxazyme G2 liquid, on milk production in pasture fed dairy cows. Anim Feed Sci Technol 145:194–208. https://doi.org/10.1016/j.anifeedsci.2007.05.049

Mohamed D, Borhami B, El-Shazly K, Sallam S (2013) Effect of supplementation with fibrolytic enzymes on the productive performance of early lactating dairy cows. J Agric Sci 5(6):146–155

Morgavi DP, Beauchemin KA, Nsereko VL, Rode LM, McAllister TA, Iwaasa AD, Wang Y, Yang WZ (2001) Resistance of feed enzymes to proteolytic inactivation by rumen microorganisms and gastrointestinal proteases. J Anim Sci 79:1621–1630. https://doi.org/10.2527/2001.7961621x

nutriNews (2020) Use of exogenous enzymes in the feeding of dairy cows. Available from: https://nutrinews.com/en/use-of-exogenous-enzymes-in-the-feeding-of-dairy-cows/?reload=yes?reload=yes?reload=yes?reload=yes?reload=yes?reload=yes?reload=yes?reload=yes. Accessed 24 Jun 2022

Ojha BK, Singh PK, Shrivastava N (2019) Enzymes in the animal feed industry. Enzyme Food Biotechnol 2019:93–109. https://doi.org/10.1016/b978-0-12-813280-7.00007-4

Pinos-Rodríguez JM, González S, Mendoza G, García JC, Miranda L, De la Cruz GA, De Lerma V (2005) Efecto de enzimas fibrolíticas exógenas en la degradación in vitro de ingredientes alimenticios y en la producción de leche de vacas Holstein. INCI 30(12):752–757

Ravindran V (2013) Feed enzymes: the science, practice, and metabolic realities. J Appl Poult Res 22(3):628–636

Reddish MA, Kung L Jr (2007) The effect of feeding a dry enzyme mixture with fibrolytic activity on the performance of lactating cows and digestibility of a diet for sheep. J Dairy Sci 90:4724–4729

Rode LM, Yang WZ, Beauchemin KA (1999) Fibrolytic enzyme supplements for dairy cows in early lactation. J Dairy Sci 82:2121–2212

SAS (2004) Sistema de análisis estadístico, guía del usuario. estadística. Versión (7.a edición), SAS. Inst. Inc. Cary, Carolina del Norte. EE. UU

Selle PH, Ravindran V (2007) Microbial phytase in poultry nutrition. Anim Feed Sci Technol 135:1–41

Sujani S, Seresinhe RT (2015) Exogenous enzymes in ruminant nutrition: a review. Asian J Anim Sci 9(3):85–99. https://doi.org/10.3923/ajas.2015.85.99

Sutton JD, Phipps RH, Beever DE, Humphries DJ, Hartnell GF, Vicini JL, Hard DL (2003) Effect of method of application of a fibrolytic enzyme product on digestive processes and milk production in Holstein-Friesian cows. J Dairy Sci 86:546–556

The Association of American Feed Control Officials (2002) Official publication. 464. Available from: https://www.aafco.org/Publications. Accessed 16 Sept 2021

Tirado-González DN, Miranda-Romero LA, Ruíz-Flores A, Medina-Cuéllar SE, Ramírez-Valverde R, Tirado-Estrada G (2018) Meta-analysis: effects of exogenous fibrolytic enzymes in ruminant diets. J Appl Anim Res 46(1):771–783. https://doi.org/10.1080/09712119.2017.1399135

Titi H, Lubbadeh WF (2004) Effect of feeding cellulase enzyme on productive responses of pregnant and lactating ewes and goats. Small Rumin Rese 52:137–143

Wang Y, Spratling BM, ZoBell DR, Wiedmeier RD, McAllister TA (2004) Effect of alkali pretreatment of wheat straw on the efficacy of exogenous fibrolytic enzymes. J Anim Sci 82:198–208. https://doi.org/10.2527/2004.821198x

Wu YB, Ravindran V, Thomas DG, Birtles MJ, Hendriks WH (2004) Influence of phytase and xylanase, individually or in combination, on performance, apparent metabolizable energy, digestive tract measurements and gut morphology in broilers fed wheat-based diets containing adequate level of phosphorus. Br Poult Sci 45:76–84

Yang WZ, Beauchemin KA, Rode LM (1999) Effects of an enzyme feed additive on extent of digestion and milk production of lactating dairy cows. J Dairy Sci 82:391–403

Yang WZ, Beauchemin KA, Rode LM (2000) A comparison of methods of adding fibrolytic enzymes to lactating cow diets. J Dairy Sci 83:2512–2520

Yang HE, Son YS, Beauchemin KA (2011) Effects of exogenous enzymes on ruminal fermentation and degradability of alfalfa hay and rice straw. Asian-Aust J Anim Sci 24(1):56–64. https://doi.org/10.5713/ajas.2011.90369

Yu P, McKinnon J, Christensen DA (2005) Improving the nutritional value of oat hulls for ruminant animals with pretreatment of a multienzyme cocktail: in vitro studies. J Anim Sci 83:1133–1141. https://doi.org/10.2527/2005.8351133x

Zhang YHP, Lynd LR (2004) Toward an aggregated understanding of enzymatic hydrolysis of cellulose: noncomplexed cellulase systems. Biotechnol Bioeng 88(7):797–824. https://doi.org/10.1002/bit.20282

Zheng W, Schingoethe DJ, Stegeman GA, Hippen AR, Treacher RJ (2000) Determination of when during the lactation cycle to start feeding a cellulase and xylanase enzyme mixture to dairy cows. J Dairy Sci 83:2319–2325

Zilio EMC, Del Valle TA, Ghizzi LG, Takiya CS, Dias MSS, Nunes AT, Silva GG, Rennó FP (2019) Effects of exogenous fibrolytic and amylolytic enzymes on ruminal fermentation and performance of mid-lactation dairy cows. J Dairy Sci 102(5):4179–4189. https://doi.org/10.3168/jds.2018-14949

Determining the Effect of Enzyme Addition to Locally Available Forages in Mongolia Using In Vitro and In Vivo Techniques

Norovsambuu Togtokhbayar, Tsevegmed Munkhnasan, Ayushjav Otgonjargal, and Choinzon Sodnomtseren

Abstract The close association between rumen fermentation and gas production has been recognized for over a century. The expansion of interest in the efficient utilization of roughage diets has led to an increase in the use of this technique worldwide, as it offers due advantage in studying fermentation kinetics of feeds during 1990s. Like the global tendency, since 2000, in Mongolia, also in vitro gas production techniques started to receive attention to improve its highly nature dependent livestock feeding system. Effects of xylanase on in vitro gas production (GP) and in sacco degradability of wheat straw was determined in laboratory and the in sacco rate of dry matter degradation decreased linearly with increasing enzymes. Intermediate levels of xylanase improved rumen kinetic fermentation and degradability. The outcome of this research indicated that the application of xylanase enzyme could improve in vitro fermentation and GP of wheat straw. In a follow-up in vivo study that tested effects of cellulase and xylanase enzyme addition into animal feed, cellulase addition was effective in improving digestibility of nutrients, growth performance, and net revenue in growing lambs. More recent studies revealed that treatment of barley straw with cellulase and xylanase enzymes showed positive effects on feed intake, digestibility, and daily body weight gain. Enzyme addition into feedstuffs holds a promise in increasing locally available forage utilization and improving the production efficiency of ruminants. Further research need to focus on application of enzymes in animal feeds and feeding management by improving both the supply of commercial enzymes in Mongolia and should consider creating conducive policy environment for institutional body to develop and produce enzyme treated feedstuffs commercially.

N. Togtokhbayar (✉) · A. Otgonjargal · C. Sodnomtseren
Mongolian University of Life Science, Ulaanbaatar, Mongolia
e-mail: n.togtokhbayar@muls.edu.mn

T. Munkhnasan
Mongolian Ministry of Food, Agriculture and Light Industry, Ulaanbaatar, Mongolia

A. Z. M. Salem et al. (eds.), *Exogenous Enzymes as Feed Additives in Ruminants*, https://doi.org/10.1007/978-3-031-27993-5_7

1 Introduction

Mongolia has the largest area of common pasture land in the world (Upton 2010), with identical short (about 90 days) growing season of pasture vegetation (Togtokhbayar 2006; Nergui et al. 2011; Otgonjargal 2012). All types of livestock are kept outdoors and the vegetation of natural pastures is the main source of feed throughout the year including winter (Munkhnasan 2016). The nutritive value of summer and autumn pasture is good, but it declines Sharply in winter and spring; therefore, winter and spring pastures provide only about half of the nutrient requirements of the animals (Nergui et al. 2011). This low and seasonal yield of native pastures causes a feed gap between the annual pasture provision and the requirements of grazing ruminants. Bynie et al. (2012) reported that Mongolian sheep lose 20.6–34.0% of their live weight during the winter and spring due to lack of supplemental feeding. Therefore, wheat and barley straws are commonly used as survival feeds for animals during the long winters when other types of feeds are short in supply.

Gendaram (1977) observed that the weight of Mongolian lambs, grazing on natural pasture, continuously declined from November to April. Traditionally, only a small number of supplements consisting of hay, cereal straws and bran, and cracked and mixed grains are offered during the winter and spring seasons and only to weak animals (Batjargal et al. 2014). Because the country does not produce sufficient quantities of hay and other feeds for supplemental feeding, wheat and barley straw have become a valuable, low cost roughage source that can be used effectively, especially in extensive ruminant production systems based on low inputs (Lopez et al. 2005).

Mongolian National Statistical Office (2015) recorded that more than 8000 tons of cereal straws were used nationwide for animal feeding in 2014. However, the availability of nutrients from these straws is much lower than the energy potentially stored in these feedstuffs. The NRC (2001) determined that typically, wheat straw contains more than 70% neutral detergent fiber (NDF) on a dry matter (DM) basis, with less than 40% total digestive tract digestibility of the NDF, even under ideal feeding conditions. Thus, a number of physical, chemical, and enzymatic treatments are used to upgrade the nutritive value of straws by increasing their digestibility through structural modifications of the polysaccharide-lignin crosslinks of the cell wall (Lopez et al. 2005).

Increasing efficiency of feed utilization has many advantages, including reducing the amount of undigested nutrients, reduction of nitrogenous compound excretion (Kuelling et al. 2003), and reduction of methane emissions (Johnson and Johnson 1995), which can be achieved by manipulating animal diets using different feed additives. Biologically active enzymes, as animal feed additives, are now produced at low cost due to recent advances in fermentation technology and biotechnology.

Exogenous fibrolytic enzymes (EFEs) may improve the nutritive value of agricultural by-products due to enhanced attachment by rumen microorganisms (Nsereko et al. 2002), creation of stable enzyme feed complexes (Kung et al. 2000), and alteration of fiber structure, which stimulates microbial colonization (Giraldo et al. 2004). Besides, exogenous fibrolytic enzymes can cause release of reducing sugar, which can enhance ruminal microbial colonization (Alsersy et al. 2015).

Several studies with EFE have reported increased microbial activities in the rumen, which resulted in enhancement of animal performance (Gado et al. 2009; Khattab et al. 2011; Salem et al. 2013; Togtokhbayar et al. 2017. However, despite the importance of enzyme formulation, the relationship between enzymatic activities and improvements in forage utilization is not well established (Eun et al. 2007). Results with EFE addition in ruminant systems have been highly variable and inconsistent, with some studies showing substantial improvements of feed digestibility and animal performance (Salem et al. 2013), while others have reported negative effects or none (Baloyi 2008). Some of the reasons postulated for the inconsistent results are that the effects of EFE are influenced by factors such as enzyme source, the type and dose of enzyme, composition of diet, enzyme application method, and even the level of animal productivity (Beauchemin et al. 2003; Carro et al. 2007).

2 Effect of Cellulase, Xylanase, and Their Combinations on In Vitro Gas Production (GP) and In Sacco Degradability of Wheat Straw

Rumen fermentation enables ruminants to digest coarse plant material while producing methane and carbon dioxide as by-products. Therefore, if the digestibility of fibrous feedstuffs in ruminant animals can be increased, then methane production could be reduced and the combined effect would reduce environmental pollution as well as improve the livestock productivity, thus improving the livelihoods of the livestock farmers (Stewart et al. 1997; Giraldo et al. 2007b). Effects of feed enzymes on in vitro gas production, *in sacco* DM degradability, and on animal performance in Mongolia have not been documented so far.

Togtokhbayar et al. (2015) conducted a research adding xylanase at doses from 0.5 to 2.0 μl/g to wheat straw (*Triticum aestivum* var. 106), that contained (g/kg on dry matter basis) crude protein, 40; ether extract, 6; neutral detergent fiber (NDF), 788; acid detergent fiber (ADF), 473; and ash, 59. The pH ranged from 6.63 to 6.74 and did not differ among treatments when it was incubated at various periods up to 48 h.

The pH decreased linearly with increasing levels of xylanase at 12 and 24 h of incubation. The lowest pH values were at intermediate enzyme levels (i.e., 1.0 to 1.5 μl/g) at 3 and 6 h of incubation (Table 1). Total GP increased quadratically with increasing levels of xylanase. The amount of gas produced with xylanase at all incubation times may be related to internal plant cell wall structures of wheat straw. It has been reported by Wallace et al. (2001) that some commercial EFE increased total GP and rates of in vitro fermentation of feed. Increased cell wall content was considered to suppress microbial activity through a reduction in the availability of rapidly fermented carbohydrates (Wilson and Hatfield 1997). However, it has been shown that EFE could potentially improve fiber degradation through a hydrolytic action prior to feeding or in vitro incubation (Giraldo et al. 2004; Elghandour et al. 2013). The increase in GP of xylanase-treated wheat straw at low and moderate levels, at all the incubation times, shows that a suitable enzyme dose improves fermentation efficiency, and/or that xylanase addition stimulated fermentation (Nsereko et al. 2002). Increased dose of xylanase may have prevented binding of enzymes to substrate receptors, which reduced proportional attachment by ruminal microorganisms to fiber (Treacher and Hunt 1996). Colombatto et al. (2003a) concluded that increasing the level of enzyme from 1× to 0.5× increased the rate and extent of GP, but levels of 10× levels were ineffective. Some studies reported that the fibrolytic enzyme did not affect GP (Giraldo et al. 2007a; Ruiz et al. 2013).

Oppositely to the GP, the ruminal volatile fatty acid (VFA) concentrations increased with low level of xylanase which is consistent with Wang and McAllister (2002) and Giraldo et al. (2005) who showed that treatment of wheat straw with an enzyme preparation containing xylanase and β-gluconase activities increased VFA at 4 and 6 h of incubation. Increase of VFA may have been due to increased fiber digestion and altered ruminal fermentation (Nsereko et al. 2002), enhanced attachment and colonization to the plant cell wall material by rumen microorganisms (Wang et al. 2001; Nsereko et al. 2002). Enzyme preparation can contain sugars that are fermented by ruminal microorganisms, thus increasing VFA production (Colombatto et al. 2003b). This seems to be dependent on many factors such as

Table 1 Effect of enzyme types, levels, and application methods on ruminal pH, total gas production, and methane emission of wheat straw at different times of incubation

Type of enzymes	Method of application	Application level	Response[a]	Reference
Xylanase	Sprayed to forage 16 h before feeding	0.5, 1.0, 1.5, 2.0 μl/g	ADF, CP, DM, EE, NDF, VFA, NH3-N	Togtokhbayar et al. (2015)
Exogenous	Added directly on the substrate	0, 5, 10, 20 μl/g	GP, DMD, ME, SCFA	Lopez et al. (2013)

[a]The type of response need to be elaborated. Is it an increase or a decrease?

ADF acid detergent fiber, *CP* crude protein, *DM* dry matter, *EE* energy efficiency, *NDF* neutral detergent fiber, *VFA* volatile fatty acid, *NH3-N* ammoniacal nitrogen, *DMD* dry matter digestibility, *ME* metabolizable energy, *SCFA* short chain fatty acid

source, type and dose of enzyme, type of diets fed to the animals and enzyme application methods, and method of administration (Beauchemin et al. 2003; Carro et al. 2007), causing inconsistent results.

The in sacco rate of DM degradation of wheat straw decreased with increasing enzyme levels, but the effective degradability values were not affected. In sacco degradation of NDF was not affected by xylanase levels (Togtokhbayar et al. 2015).

Xylanase enzyme increased both potentially degradable fraction and degradation rate of NDF and ADF. The positive results of enzymes on in vitro degradation of substrate fiber are consistent with those reported by Yang et al. (2000) and Salem et al. (2015a, b). The mode of action by which enzymes can affect digestion is still subject to speculation (Elwakeel et al. 2007). It has been suggested that improved degradation could be related to the fact that enzymes treatments increase bacterial colonization of feed particles (Yang et al. 2000), and reduce the release of sugar-digested portion of the feed-attracted bacteria to the site of digestion and enhance attachment of bacteria to undigested feed particles (Yang et al. 2000; Wang et al. 2001). The solubilized carbohydrates would also provide energy that would lead to rapid microbial growth, shortening the lag time for microbial colonization (Yang et al. 2000). Dawson and Tricarico (1999) suggested that the most active period for enzyme effect is the first 6–12 h of the digestive process, prior to bacterial colonization of feed substrates.

Greater effects with low or moderate amounts of xylanase than greater amounts of this enzyme agree with the results reported by Beauchemin et al. (2000). Furthermore, Beauchemin et al. (2003) speculated that excessive enzymes applied to the feed may bind to sites used by rumen bacteria and make them unavailable, creating a barrier against microbial colonization.

Newbold (1997) noted that most enzymes function within a few hours of feeding before being degraded by the proteolytic activity of rumen microbes. In our study, applying xylanase to wheat straw increased, linearly, both the degradation rate and extent of DM and NDF in the rumen only at 3, 6, and 12 h after incubation. These results suggest that the treatment with xylanase stimulated the initial phases of feed degradation, but the effects were reduced as incubation time progressed.

Similarly, in our study, the optimal dose for the degradability of wheat straw was intermediate dosage (1.0–1.5 μl/g) compared with the low or high levels of enzyme addition.

3 Effects of Exogenous Cellulase and Xylanase Enzyme Preparations on Feed Intake, Nutrient Digestibility, Growth, and Economics of Rearing Mongolian Lamb

The use of exogenous enzyme preparations in ruminant diets is an efficient technology for improvement of forage quality and has gained lot of interest in recent years (Beauchemin et al. 2003; Salem et al. 2015a, b). It is well known that supplementing animal rations with fiber-degrading enzymes can improve feed utilization by enhancing fiber degradation in vitro (Table 1;Diaz et al. 2013; Salem et al. 2015a, b; Togtokhbayar et al. 2015) and animal performance (Gado et al. 2009; Khattab et al. 2011; Salem et al. 2013; Alsersy et al. 2015; Rojo et al. 2015; Valdes et al. 2015).

3.1 Feed Intake and Digestibility

Significant effects were observed for the intake of nutrients with diet B+ENZc /cellulase enzyme–treated barley straw/ and for the crude protein (CP) and NDF digestibility of diet W+ENZc /cellulase enzyme–treated wheat straw/, whereas no effects were found for xylanase addition (Table 2). The DM, CP, and NDF intake of B+ENZc increased by 10.9%, 5.7%, and 16.6%, respectively, in comparison to B+ENZ (Table 2). Compared to diet W+ENZc, the CP intake was 7.5% lower ($p \leq 0.05$) and NDF intake was 22.9% higher ($p \leq 0.05$). The DM and organic matter (OM) intake of B+ENZc were higher ($p \leq 0.05$) than of B+ENZx /xylanase enzyme–treated barley straw/.

Effects on nutrient intake and digestibility from the study done by Togtokhbayar et al. (2015) were similar to those reported by Hussain et al. (2008) and Meale et al. (2014) who concluded that the application of exogenous enzyme preparations to the diets of small ruminants has little impact on nutrient intake and production. However, the application of exogenous enzyme preparations had a positive effect on body weight gain and was therefore in agreement with findings of in vivo studies with cattle (Feng et al. 1996; Beauchemin et al. 1999) and lambs (Giraldo et al. 2008) who reported that enzyme supplementation of the daily diet resulted in a higher average daily gain.

In general, the digestibility of the various proximate constituents was higher for diet W+ENZc than for diet B+ENZc. This is in agreement with Wang and McAllister (2002) and Jafari et al. (2004) who also found that wheat straw treated with a fibrolytic enzyme enhanced CP, NDF, and ADF digestibility (in vivo). Also Lewis et al. (1999) and Rode et al. (1999) reported that total digestibility of nutrients was increased by enzyme treatment and that supplementing dairy cow diets with a fibrolytic enzyme mixture has the potential to enhance milk yield and nutrient digestibility of cows in early lactation without affecting feed intake.

3.2 Growth Performance and Feed Efficiency

In general, the addition of cellulase and xylanase enzyme preparations increased the average daily gain of lambs by 60–102% and 32–48%, respectively, compared to their control groups (Table 2). This was probably due to the improved nutrient digestibility and thus enhanced feed efficiency. Kung et al. (2000) noticed that treatment of feeds with enzyme preparations just prior to feeding can improve

Table 2 Effect of enzyme types, levels, and application methods on dry matter intake (DMI), average daily gain (ADG), and feed convention rate (FCR)

Type of enzymes	Method of application	Application level	Forage level	Effect DMI	ADG	FCR	Total	Reference
Xylanase	Sprayed to forage 16 h before feeding	1.0 μl/g	60%	–	–	–	CP, DM, EE, NDF,	Togtokhbayar et al. (2017)
Fibrolytic	Enzymes diluted in 240 ml water and sprayed to forage	5.0, 7.5 ml/ kg DM	60%	–	–	–	DM, NDF, ADF, NH_3-N, VFA,	Torres et al. (2013)
Exogenous glucoamylase		1.5 ml/ kg DM	45%	–	–	–	DM, NDF, CP, DM, EE	Mendoza et al. (2013)
Fibrolytic	Enzymes were diluted in 300 ml water and sprayed to forage	3.5 ml kg DM	60%	–	–	*	DM, NDF, ADF, NH_3-N, VFA,	Torres et al. (2013)
Exogenous	Basal diet 10 g of enzyme	10 g/TMR	70% corn silage, 30% TMR	*	*	*		Cayetano et al. (2013)
Lignocellulolytic	Sprayed to forage 24 h before feeding	60, 120 ml/ kg DM	–	–	–	*	DM, CP, NDF, ADF,	Arce-Cervantes et al. (2013)
Exogenous fibrolytic	Sprayed to directly on the oat straw	5, 10 g/kg^{-1} DM	60% oat straw	–	–	*	DM, NDF, ADF, NH_3-N, VFA,	Lara Bueno et al. (2013)
Exogenous	Basal diet 10 g of enzyme	10 g/TMR	70% corn silage, 30% TMR	–	–	*		Rivero et al. (2013)

ADF acid detergent fiber, *CP* crude protein, *DM* dry matter, *EE* energy efficiency, *NDF* neutral detergent fiber, *VFA* volatile fatty acid, *NH_3*-N ammoniacal nitrogen, $p<0.05$

digestibility via different mechanisms: enhanced microbial attachment, changes in gut viscosity, complementary actions with ruminal enzymes, changes in the site of nutrient digestion, improvements in palatability, and changes in the patterns of feed consumption.

These results are consistent with findings of other authors (Beauchemin et al. 1997; Lewis et al. 1999; Titi and Lubbadeh 2003; Salem et al. 2011) who suggested that the improved animal performance through feeding of enzyme treated straw was due to increased digestibility, which enhances energy and or nutrient availability to rumen microbes. However, other studies in sheep (McAllister et al. 2000; Miller et al. 2008) and cattle (Krueger et al. 2008) reported no effects of enzyme treatments applied at feeding on average daily gain. Although the results of using fibrolytic enzyme preparations in lambs' rations in the current study are encouraging in general, the results of such applications in animal trials have been inconsistent due to varying enzyme doses, different ways of enzyme preparations, different feed rations, and animal species. Using the cellulase enzyme preparation improved the feed conversion ratio of diet with wheat straw and diet with barley straw by 30% and 45% over the diet without enzyme additions, and the xylanase treatment increased the feed conversion ratio of diet wheat straw and barley straw by 22% and 31%, respectively, which is in agreement with Brossard et al. (2006). The improved feed conversion ratio observed in the study by Holtshausen et al. (2011) and Togtokhbayar et al. (2017) might be attributable to the enhanced NDF digestibility.

3.3 Economic Evaluation

Enzyme application had no effect on the total feed cost for both diet types but in both cases showed positive effects on daily gain, net revenue, economic, and relative economic feed efficiency (Togtokhbayar et al. 2017). The cellulase enzyme treatment showed the highest improvements in the above parameters compared to the xylanase treatment. Cellulase- and xylanase-treated barley straw showed numerically higher body weight gain, price for body weight gain, net revenue, and feed efficiency than diet W (wheat straw + wheat bran), but differences were only significant for net revenue and relative feed efficiency (Table 3).

For both diets, calculated net revenue and economic feed efficiency were substantially higher for both enzyme treatments as compared to the control diets. The cellulase enzyme addition showed higher positive effect on net revenue and feed efficiency in comparison to the xylanase treatment. Without enzyme treatment, diet B (barley straw + wheat bran) yielded lower values than diet W for all economic variables (Table 3). In short, studies conducted in Mongolia suggested that economically, the cellulase enzyme preparation performs better than the xylanase preparation for both types of diets. These studies agree with those of Tozer et al. (2003) who found that although costs per kilogram of milk produced were lowest for pasture-concentrate

Table 3 Economic evaluation of feeding Mongolian lambs without (–ENZ) or with cellulase-treated (+ENZc) or xylanase-treated (+ENZx) diet W (wheat straw + wheat bran) and diet B (barley straw + wheat bran)

Variable	Diet	-ENZ	+ENZc	+ENZx	SEM	$P =$
Total feed costs (MNT)	W	7780	8664	8100	237.1	0.348
	B	7749	8631	7883	173.2	0.065
Body weight gain (kg)	W	2.3^{c}	3.6^{a}	3.0^{b}	0.2	0.001
	B	2.1^{c}	4.2^{a}	3.1^{b}	0.3	0.001
Price of body weight gain (MNT)	W	15867^{c}	25200^{a}	20767^{b}	1416.2	0.001
	B	14525^{c}	29225^{a}	24525	1975.1	0.001
Net revenue (MNT)	W	8086^{c}	16536^{Ba}	12665^{b}	1293.5	0.001
	B	6776^{c}	20595^{Aa}	13642^{b}	1858.6	0.000
Economic feed efficiency (%)	W	104^{b}	193^{a}	157^{ab}	8.3	0.013
	B	88.6^{b}	238^{a}	173^{a}	21.0	0.001
Relative economic efficiency (%)	W	100^{b}	186^{Ba}	$151^{B\ ab}$	15.0	0.006
	B	100^{a}	268^{Aa}	$195^{B\ b}$	12.9	0.001

Source: Togtokhbayar et al. (2017)
W and B represent diet W (wheat straw + wheat bran) and diet B (barley straw + whaet bran), respectively. SEM is standard error of means; *p* is probability of a variable effect
A, *B*: Capital letters within one column indicate significant difference between the two diets
a, *b*, *c*: Superscripts within one row indicate significant differences within a diet treated with different exogenous enzyme preparations and/or untreated diet

feeding, an enzyme-treated total mixed ration yielded the highest net income per cow per day because of higher yields of milk and milk components.

4 Conclusions

The treatment of wheat and barley straw with the cellulase enzyme improved the digestibility of organic matter, crude protein and NDF, increased average daily body weight gain, feed efficiency, and net revenue as compared to untreated control diets. The higher economic feed efficiency with low feed cost per kg live weight gain of lambs fed with the cellulase enzyme–treated wheat and barley straws in this chapter suggest economic feasibility of this enzyme preparation and point to its potential for the production of good-quality winter feed for ruminants in Mongolia.

However, application of enzymes in animal feed remains in research stage and not in use in animal feeding management in Mongolia due to low availability of the commercial enzymes and no institutional body to develop an enzyme to be used for treating animal feedstuffs. Further studies are highly recommended to consider how to improve the nutritional quality of feedstuffs treated with commercial enzymes.

Acknowledgments This study was conducted under the IAEA Co-ordinated Research Project D3.10.27. Research Contract #MON16296. The authors thank the IAEA for financial and technical support.

References

Alsersy H, Salem AZM, Borhami BE, Olivares J, Gad HM, Mariezcurrena MD, Yacuot MH, Kholif AE, El-Adawy M, Hernandez SR (2015) Effect of Mediterranean saltbush (Atriplex halimus) ensilaging with two developed enzyme cocktails on feed intake, nutrient digestibility and ruminal fermentation in sheep. Anim Sci J 85:51–58

Arce-Cervantes O, Mendoza GD, Hernbndez PA, Meneses N, Torres Salado N, Loera (2013) The effects of a lgnocellulotc extract of Fomes sp. EUM1 on the intake. Digestibility, feed efficiency and growth of lambs. In: Salem AZM, Odongo N (eds) Animal nutrition and feed technology, special issue entitled exogenous enzymes in animal nutrition-benefits and limitations, pp 369–372. Izatnagar-243 122, India

Baloyi TF (2008) Effects of exogenous fibrolytic enzymes on in vitro fermentation kinetics of forage and mixed feed substrates. Dissertation, Stellenbosch University

Batjargal N, Togtokhbayar N, Tserendash S (2014) Final report on animal feed resources and their management in Mongolia. In: FAO regional meeting titled dairy Asia towards sustainability, Bangkok, Thailand, 21–24 May 2014

Beauchemin KA, Jone SD, Rode LM, Sewalt VJ (1997) Effects of fibrolytic enzymes in corn or barley diets on performance and carcass characteristics of feedlot cattle. Can J Anim Sci 77:243–246

Beauchemin KA, Rode LM, Karren D (1999) Use of feed enzymes in feedlot finishing diets. Can J Anim Sci 79:243–246

Beauchemin KA, Rode LM, Maekawa M, Morgavi DP, Kampen R (2000) Evaluation of a nonstarch polysaccharidase feed enzyme in dairy cow diets. J Dairy Sci 83:543–553

Beauchemin KA, Colombatto D, Morgavi DP, Yan WZ (2003) Use of exogenous Fibrolytic enzymes to improve feed utilization by ruminants. J Anim Sci 81:37–47

Brossard L, Chaucheyras DF, Michalet DB, Martin C (2006) Dose effect of live yeasts on rumen microbial communities and fermentations during butyric latent acidosis in sheep: new type of interaction. J Anim Sci 82:1–11

Bynie B, Danjin J, Choijil D (2012) Barga breed sheep. Ulaanbaatar, Mongolia

Carro MD, Giraldo LA, Mohamed AH, Ranilla MJ, Tejido ML (2007) In vitro ruminal fermentation of low-quality forages as influenced by the treatment with exogenous fibrolytic enzymes. In: Priolo A, Biondi L, Ben Salem H, MorandFehr P (eds) Advanced nutrition and feeding strategies to improve sheep and goat. CIHEAM, Zaragoza, pp 263–267

Cayetano JA, Salem AZM, Marezcurrena BMA, Roji R, Cerrllo-Soto MA, Gado H, Camacho LM (2013) Effect of adding Salix babylonica extracts and exogenous enzymes to basal diets on the meat quality of growing Suffolk lambs. In: Salem AZM, Odongo N (eds), Animal nutrition and feed technology, special issue entitled exogenous enzymes in animal nutrition-benefits and limitations. pp. 373–380. Izatnagar-243 122, India

Colombatto D, Mould FL, Bhat MK, Morgavi DP, Beauchemin KA, Owen E (2003a) Influence of fibrolytic enzymes on the hydrolysis and fermentation of pure cellulose and xylan by mixed ruminal microorganisms in vitro. J Anim Sci 81:1040–1050

Colombatto D, Mould FL, Bath MK, Morgavi DP, Owen E (2003b) Use of fibrolytic enzymes to improve the nutritive value of ruminant diets. A biochemical and in vitro rumen degradation assessment. Anim Feed Sci Technol 107:201–209

Dawson KA, Tricarico JM (1999) The use of exogenous fibrolytic enzymes in ruminants. In: Jacques K, Lyons TP (eds) 15th annual symposium on biotechnology in the feed industry. Nottingham University Press, Nottingham, pp 303–312

Diaz A, Carro MD, Saro C, Mateos I, Odongo EN (2013) In vitro evaluation of commercial fibrolytic enzymes for improving the nutritive value of low-quality forages. Anim Nutr Feed Technol 13:461–476

Elghandour MMY, Salem AZM, Ronquillo M, Bórquez JL, Gado HM, Odongo NE, Peñuelas CG (2013) Effect of exogenous enzymes on in vitro gas production kinetics and ruminal fermentation of four fibrous feeds. Anim Feed Sci Technol 179:46–53

Elwakeel EA, Titgemeyer EC, Johnson BJ, Armendariz CK, Shirley JE (2007) Fibrolytic enzymes to increase the nutritive value of dairy feedstuffs. J Dairy Sci 90:5226–5236

Eun JS, Beauchemin KA, Schulze H (2007) Use of exogenous fibrolytic enzymes to enhance in vitro fermentation of alfalfa hay and corn silage. J Dairy Sci 90:1440–1451

Feng P, Hunt CW, Pritchard GT, Julien WE (1996) Effect of enzyme preparations on in situ and in vitro degradation and in vivo digestive characteristics of mature cool-season grass forage in beef steers. J Anim Sci 74:1349–1357

Gado HM, Salem AZM, Robinson PH, Hassan M (2009) Influence of exogenous enzymes on nutrient digestibility, extent of ruminal fermentation as well as milk production and composition in dairy cows. Anim Feed Sci Technol 154:36–46

Gendaram K, (1977) Stipa- Cleistogenes pasture yield, chemical composition and nutrient dynamics in the steppe zone of Mongolia

Giraldo LA, Ranilla MJ, Tejido ML, Carro MD (2004) Effects of enzyme application method on the in vitro rumen fermentation of tropical forages. J Anim Feed Sci 13:63–66

Giraldo LA, Carro MD, Ranilla MJ, Tejido ML (2005) Influence of fibrolytic enzymes on in vitro methane production and rumen fermentation of a substrate containing 60% of grass hay. In: 11th seminar of the FAO-CIHEAM subnetwork on sheep and goat nutrition, Catania, Italy, 8–10 Sep 2005

Giraldo LA, Carro MD, Ranilla MJ, Tejido ML (2007a) Influence of fibrolytic enzymes on in vitro methane production and rumen fermentation of a substrate containing 60% of grass hay. In: Priolo A, Biondi L, Ben Salem H, Moran d-Fehr P (eds) Advanced nutrition and feeding strategies to improve sheep and goat. CIHEAM, Zaragoza, pp 257–261

Giraldo LA, Tejido ML, Ranilla MJ, Carro MD (2007b) Effects of exogenous cellulase supplementation on microbial growth and ruminal fermentation of a high-forage diet in Rusitec fermenters. J Anim Sci 8:1962–1970

Giraldo LA, Tejido ML, Ranilla MJ, Ramos S, Carro MD (2008) Influence of direct-fed fibrolytic enzymes on diet digestibility and ruminal activity in sheep fed a grass hay-based diet. J Anim Sci 86:1617–1623

Holtshausen L, Chung YH, Gerardo-Cuervo H, Oba M, Beauchemin KA (2011) Improved milk production efficiency in early lactation dairy cattle with dietary addition of a developmental fibrolytic enzyme additive. J Dairy Sci 94:899–907

Hussain A, Nisa M, Sarwar M, Sharif M, Javaid A (2008) Effect of exogenous fibrolytic enzymes on ruminant performance. Pak J Agric Sci 45:297–306

Jafari A, Alikhani M, Edriss MA (2004) The effect of substituting treated wheat straw with fibrolytic enzymes for alfalfa hay on Naieni replacement ewe lambs performance. In: Proceeding of 11th AAAP congress, Volume III. Malaysia, pp 610–613

Johnson KA, Johnson DE (1995) Methane productions from cattle. J Anim Sci 73:2483–2492

Khattab HM, Gado HM, Kholif AE, Mansour AM, Kholif AM (2011) The potential of feeding goats sun dried rumen contents with or without bacterial inoculums as replacement for berseem clover and the effects on milk production and animal health. Int J Dairy Sci 6:267–277

Krueger NA, Adesogan AT, Staples CR, Krueger WK, Kim SC, Littell RC, Sollenberger LE (2008) Effect of method of applying fibrolytic enzymes or ammonia to Bermudagrass hay on feed intake, digestion and growth of beef steers. J Anim Sci 86:882–889

Kuelling DR, Menzi H, Sutter F, Lischer P, Kreuzer M (2003) Ammonia, nitrous oxide and methane productions from differently stored dairy manure derived from grass- and hay-based rations. Nut Cycl Agroecosys 65:13–22

Kung L, Treacher RJ, Nauman GA, Smagala AM, Endres KM, Cohen MA (2000) The effect of treating forages with fibrolitic enzymes on its nutritive value and lactation performance of dairy cows. J Dairy Sci 83:115–122

Lara Bueno A, Mendoza Martinez GD, Hernandez Garcia PA, Martinez Garcia GD, Plata Perez FX (2013) Evaluation of high doses of exogenous fibrolytc enzymes in lambs fed an oat straw based raton. In: Salem AZM, Odongo N (eds), Animal nutrition and feed technology, special

issue entitled exogenous enzymes in animal nutrition-benefits and limitations. pp. 355–362. Izatnagar-243 122, India
Lewis GE, Sanchez WK, Hunt CW, Guy MA, Pritchard GT, Swanson BJ, Trearcher RJ (1999) Effect of direct fed fibrolitic enzymes on the lactational performance of dairy cows. J Dairy Sci 82:611–617
Lopez S, Davies DR, Giraldez FJ, Dhanoa MS, Dijkstra J, France J (2005) Assessment of nutritive value of cereal and legume straws based on chemical composition and in vitro digestibility. J Sci Food Agr 85:1550–1557
McAllister TA, Stanford K, Bae HD, Treacher RJ, Hristov AN, Baah J, Shelford JA, Cheng KJ (2000) Effect of a surfactant and exogenous enzymes on digestibility of feed and on growth performance and carcass traits of lambs. Can J Anim Sci 80:35–44
Meale SJ, Beauchemin KA, Hristov AN, Chaves AV, McAllister TA (2014) Board-invited review: opportunities and challenges in using exogenous enzymes to improve ruminant production. J Anim Sci 92:427–442
Mendoza GD, Mota N, Plata FX, Martinez JA, Hernandez PA (2013) Effects of exogenous glucoamylase from aspergillus Niger and grain level on performance of the lambs. In: Salem AZM, Odongo N (eds), Animal nutrition and feed technology, special issue entitled exogenous enzymes in animal nutrition-benefits and limitations. pp. 391–398. Izatnagar-243 122, India
Miller DR, Elliott R, Norton BW (2008) Effects of an exogenous enzyme; Roxazyme ®G2 liquid on digestion and utilization of barley and sorghum grain based diets by ewe lambs. Anim Feed Sci Technol 140:90–109
Munkhnasan TS (2016) Herd management and livestock productivity in the Altai region of Western Mongolia. Doctoral theses, University of Kassel, Witzenhausen, Germany
National Statistical Office of Mongolia (2015) Yearly statistical book-2014 Ulaanbaatar, Mongolia. Available at: http://www.nso.mn/eng/index.php. Accessed 03 Feb 2017
Nergui D, Jigjidpurev S, Otgoo T, Dugersuren S, Lkhagvajav D, Enkhtsetseg B (eds) (2011) Animal feed and feeding research for 50th anniversary of Mongolian Institute of Animal Husbandry, Ulaanbaatar
Newbold J (1997) Proposed mechanisms for enzymes as modifiers of ruminal fermentation. Processing of 8th annual Florida ruminant nutrition symp. Gainesville, FL, USA pp. 146–159
Nsereko VL, Beauchemin KA, Morgavi DP, Rode LM, Furtado AF, McAllister TA (2002) Effect of a fibrolytic enzyme preparation from Trichoderma longibrachiatum on the rumen microbial population of dairy cows. Can J Microbiol 48:14–20
Otgonjargal A (2012) Nutritive value, methane and carbon dioxide production from pastures of mountainous steppe. Doctoral dissertation, Mongolian State University of Agriculture, Ulaanbaatar
Rivero N, Salem AZM, Ronquillo MG, Cerrillo-Soto MA, Camacho LM, Gado H, Pecuelas CG (2013) Effects of exogenous enzmes and Salix babylonica L. extract on cellular immune response and its correlation with average dail weight gain in growing lambs. In: Salem AZM, Odongo N (eds), Animal nutrition and feed technology, special issue entitled exogenous enzymes in animal nutrition-benefits and limitations. pp. 411–422. Izatnagar-243 122, India
Rode LM, Yang WZ, Beauchemin KA (1999) Fibrolytic enzyme supplements for dairy cows in early lactation. J Dairy Sci 82:2121–2126
Rojo R, Kholif AE, Salem AZM, Elghandour MMY, Odongo NE, Montes de Oca R, Rivero N, Alonso MU (2015) Influence of cellulose addition to dairy goat diets on digestion and fermentation. Milk production and fatty acids content. J Agr Sci 153(8):1514–1523
Ruiz PJA, Moreno AJ, Salem AZM, Castelan Ortega O, Gonzalez-Ronquillo M (2013) Chemical composition and in vitro gas production from different varieties of native and hybrid maize silage with the addition of acetic acid or enzymes. Anim Nut Feed Technol 13:593–599
Salem AZM, El-Adawy M, Gado H, Camacho LM, Gonzalez-Ronquillo M, Alsersy H, Borhami B (2011) Effects of exogenous enzymes on nutrients digestibility and growth performance in sheep and goats. Trop Subtrop Agroecosys 14:867–874

Salem AZM, Gado HM, Colombatto D, Eghandour MMY (2013) Effect of exogenous enzymes on nutrient digestibility, ruminal fermentation and growth performance in beef steers. Livest Sci 154:69–73

Salem AZM, Buendía-Rodríguez G, Elghandour MMY, Mariezcurrena Berasain MA, Peña Jiménez FJ, Pliego AP, Chagoyán JCV, Cerrillo MA, Rodríguez MA (2015a) Effects of cellulase and xylanase enzymes mixed with increasing doses of Salix babylonica extract on in vitro rumen gas production kinetics of a mixture of corn silage with concentrate. J Integr Agric 14:131–139

Salem AZM, Alsersy H, Camacho LM, El-Adawy MM, Elghandour MMY, Kholif AE, Rivero N, Alonso MU, Zaragoza A (2015b) Feed intake nutrient digestibility nitrogen utilization and ruminal fermentation activities in sheep fed at-riplex halimus ensiled with three developed enzyme cocktails. Czech J Anim Sci 60:185–194

Stewart CS, Flint HJ, Bryant MP (1997) The rumen bacteria. In: Hobson PN, Stewart CS (eds) The Rumen Microbial Ecosystem. Chapman & Hall, London, pp 10–72

Titi H, Lubbadeh WF (2003) Effect of cellulase enzyme on productive response of pregnant and lactating ewes and goats. Small Ruminant Res 52:137–143

Togtokhbayar N (2006) Evaluation of pasture forages as livestock feed. In: Makkar HPS, Smith T (Eds) Improving animal productivity by supplementary feeding of multinutrient blocks, controlling internal parasites and enhancing utilization of alternate feed resources. International Atomic Energy Agency (IAEA-TECDOC-1495), 211–218

Togtokhbayar N, Cerrillo MA, Rodríguez GB, Elghandour MM, Salem AZ, Urankhaich C, Jigjidpurev S, Odongo NE, Kholif AE (2015) Effect of exogenous xylanase on rumen in vitro gas production and degradability of wheat straw. J Anim Sci 86:765–771

Togtokhbayar N, Urankhaich C, Otgonjargal A, Munkhnasan TS, Odongo NE (2017) Effects of exogenous cellulase and xylanase enzyme preparations on feed intake, nutrient digestibility, growth and economics of rearing Mongolian lambs. J Agric Rural Dev Trop 118(1):81–89

Torres N, Mendoza GD, Barcena R, Loera O, Gonzalez S, Aranda E, Hernandez PA, Croby M (2013) Effects of various fibrolytic enzyme extracts on digestibility and productive performance of lambs fed a forage-based diet. In: Salem AZM, Odongo N (eds), Animal nutrition and feed technology, special issue entitled exogenous enzymes in animal nutrition-benefits and limitations. pp. 381–389. Izatnagar-243 122, India

Tozer PR, Bargo F, Muller DL (2003) Economic analyses of feeding systems combining pasture and total mixed ration. J Dairy Sci 86:808–818

Treacher RJ, Hunt CW (1996) Recent developments in feed enzymes for ruminant rations. In: Proceedings of Pacific Northwest Animal Nutrition Conference, Seattle, WA, USA p 37–54

Upton C (2010) Living off the land: nature and nomadism in Mongolia. Geoforum 41(6):865–874

Valdes KI, Salem AZM, Lopez S, Alonso MU, Rivero N, Elghandour MMY, Domínguez IA, Ronquillo MG, Kholif AE (2015) Influence of exogenous enzymes in presence of Salix babylonica extract on digestibility, microbial protein synthesis and performance of lambs fed maize silage. J Agric Sci 153(4):732–742

Wallace RJ, Wallace SJ, McKain N, Nsereko VL, Hartnell GF (2001) Influence of supplementary fibrolytic enzymes on the fermentation of corn and grass silages by mixed ruminal microorganisms in vitro. J Anim Sci 79:1905–1916

Wang Y, McAllister TA (2002) Rumen microbes enzymes and feed digestion: a review. Asian-Australas J Anim Sci 15:1659–1676

Wang Y, McAllister TA, Rode LM, Beauchemin KA, Morgavi DP, Nsereko VL (2001) Effects of an exogenous enzyme preparation on microbial protein synthesis, enzyme activity and attachment to feed in the rumen simulation technique (Rusitec). Br J Nutr 85:325–332

Wilson JR, Hatfield RD (1997) Structural and chemical changes of cell wall types during stem development: consequences for fibre degradation by rumen microflora. Australas J Agr Res 48:165–180

Yang WZ, Beauchemin KA, Rode LM (2000) A comparison of methods of adding fibrolytic enzymes to lactating cow diets. J Dairy Sci 83:2512–2520

Fungi as a Source of Exogenous Enzymes in Ruminant Feeding

Mona Mohamed Mohamed Yasseen Elghandour, Abdelfattah Zeidan Mohamed Salem, Moyosore Joseph Adegbeye, José Luis Ponce-Covarrubias, Gustavo Tirado Estrada, and Pedro Enrique Hernández Ruiz

Abstract The complex link between plant cell walls has rendered rumen microbes incapable of completely degrading feed consumed by ruminants, resulting in their inability to maximally utilize the nutrients available in the feed. In such cases, rumen microbes are unable to complete the fermentation process within the retention time of the feed in the rumen. *Pleurotus ostreatus* and *Armillariella tabescens* were found capable of degrading mycotoxin (aflatoxin) present in feed. Incubation of fodder with *P. ostreatus* and *Trametes versicolor* can improve its nutrient quality by increasing the protein content while rapidly degrading the acid detergent fiber in the plant biomass. Cellulase and xylanase enzymes from *Trichoderma* spp., *Aspergillus awomori*, and *Piromyces* fungi from the wild are capable of efficiently degrading starch, rice bran, banana peel, plant secondary metabolites, agroindustry by-products, and other crop residues and milling by-products. Fungal enzymes can be used in dairy and feedlot production systems, as tbey increased acetate and butyrate production in high fiber-containing diets and propionate in low forage rations. However, it should be noted that increased fiber digestion is sometimes associated with increased methane production. Feed treated with an appropriate enzyme or a combination of enzymes at different inclusion levels can improve ruminant nutrition especially in developing countries where feeds are scare and of poor quality and farmers face seasonal supply of forages due to an extended dry season.

M. M. M. Y. Elghandour · A. Z. M. Salem (✉)
Facultad de Medicina Veterinaria y Zootecnia, Universidad Autónoma del Estado de México, Toluca, Estado de México, Mexico

M. J. Adegbeye
Department of Animal Production and Health, Federal University of Technology, Akure, Nigeria

J. L. Ponce-Covarrubias · P. E. H. Ruiz
Escuela Superior de Medicina Veterinaria y Zootecnia No. 3, Universidad Autónoma de Guerrero, Tecpan de Galeana, Guerrero, Mexico

G. T. Estrada
Departamento de Ingenierías, Tecnológico Nacional de México (TecNM)/Instituto Tecnológico El Llano Aguascalientes, Aguascalientes, Mexico

A. Z. M. Salem et al. (eds.), *Exogenous Enzymes as Feed Additives in Ruminants*, https://doi.org/10.1007/978-3-031-27993-5_8

1 Introduction

Ruminant nutritionists in developing countries are faced with an enormous challenge of overcoming the limitations in the efficiency of digestion of many fibrous crop residues and tropical grasses. This, together with high feed costs, (Sujani and Seresinhe 2015) is a major constraint that limits the productivity of ruminant livestock in most African and Asian countries where poor forage quality and low protein diets characterize their production systems. Poor ruminant productivity under such conditions is largely due to inefficiency of microbes in fermenting the feed within the rumen, because of the complex links in the dietary cell wall fractions that limit the degradation process (Roque et al. 2017). Ruminants primarily feed on forages (grazed and/or harvested herbage and browse be it the form of green, silages, hay, straws, stover, etc.) and those reared in intensive production systems usually receive concentrates in their diets in addition to forages. The main components of most forages are cellulose, hemicellulose, and lignin. Rumen microbes and intestinal enzymes are responsible for the digestion and hydrolysis of feed ingredients, providing absorbable nutrients to the host animal for maintenance and production.

Due to increasing worldwide demand of food from ruminant livestock (beef, mutton, goat meat, and milk) and rising competition between humans and animals for food and feed resources, there is a need to improve ruminant production efficiency. Innovative research and the application of appropriate technologies would help to meet this requirement. One option would be to increase the level of grain (i.e., starch in the diet); however, there are strong arguments for use of grain in ruminant diets due to competition with humans and nonruminant livestock. Furthermore, increasing the grain contents of ruminant diets can potentially be detrimental to animal health due to acidosis and other metabolic diseases. A second option for improving ruminant productivity through dietary means would be to improve the digestibility of various feed ingredients, but some dietary supplementation may be necessary to overcome any nutrient deficiencies. Crop residues and agroindustry by-products have been widely considered as ruminant feed components, since they are potentially rich in nutrients, but the availability of these nutrients for utilization by gut microbe and the host animal can be severely limited due to the structural complexity of the cell walls of these feed items (Narra et al. 2012). Strategies to enhance digestibility such as rumen microbial manipulation, pretreatment of feed ingredients, and dietary supplementation have been tested. Some dietary supplements such as probiotics, prebiotics, antibiotics, ionophores, plant extracts, and exogenous enzymes have been studied. There is particular interest in the use of enzymes in the livestock industry to improve feed breakdown and animal performance (Guimarães et al. 2006; van den Brink and de Vries 2011; Morsy et al. 2016).

Among microbes capable of producing beneficial enzymes, filamentous fungi are of interest, because they are easy to cultivate (Guimarães et al. 2006). Apart from animal livestock industry, enzymes have been used in the agroindustry to convert substrates during brewing and used in paper production in addition to the vegetable, textile, and food industries (Østergaard and Olsen 2010). More than 25% of all commercial enzymes are produced by *Aspergillus*, while other fungi, such as *Penicillium*, *Trichoderma*, *Rhizopus*, and *Humicola*, account for another 20% (Østergaard and Olsen 2010).

Fungi produce a wide range of highly active plant-fiber-degrading enzymes, which makes them generally more effective in degrading feed materials than other microbe types. They are capable of degrading cellulose, starch, xylan and other hemicelluloses, lignin, pectin, and protein. The concentrations of enzymes produced by fungi depend on the substrates they act upon. This could be due to their more effective ability to infiltrate plant tissues than other microbes, including those found in the ruminant gut (Lowe et al. 1987). Fungi play a significant role in degrading plant remnants and other organic matter present on the ground and in the soil, because they secrete a complex battery of enzymes (Østergaard and Olsen 2010). The fiber-degrading role of fungi is exhibited in nature through the natural decomposition of dead trees, stumps, and other fibrous and lignified materials as evident by the growth of mushrooms on trees, and cultivation of edible mushrooms by inoculating sawdust or palm fruit bunch with mushroom hyphae.

In many developing countries, especially those in the tropics, most grazed and harvested forages, and other materials available for feeding to ruminants, including crop residues, and milling by-products, contain complex fiber with poor digestibility. The use of fungal enzymes has the potential to improve digestibility and enhance productivity. The problem is that these forages and by-products, which are high in fiber and lignin, tend to be resistant to breakdown by conventional fermentative activity in the rumen. Thus, the use of microbes capable of degrading lignocellulolytic substrates could overcome this problem (Dhillon et al. 2011). It is not only the structural make-up of cell walls that inhibits the ability of microbes to degrade plant biomass, but many plants also contain antinutritional factors (Dhillon et al. 2011). Veum et al. (2006) and Simons et al. (1990) suggested that adding appropriate enzymes to feeds could help in reducing antinutritional effects and, thus, potentially improve the nutritive value of forages. Whereas bacteria and fungi are responsible for breakdown of plant material in the forestomach and hind gut of the ruminant, pretreatment of fibrous feeds to hydrolyze plant cell walls by chemical means, such as with sodium hydroxide or ammonia, can also increase digestibility and nutritive value. Exogenous enzymes, especially those of fungal origin, may also be useful in enhancing the breakdown of fibrous materials. The objective of this review is to specifically evaluate the influence of enzymes produced by filamentous fungi on ruminant nutrition.

2 Exogenous Enzymes Produced by Fungi and Their Actions

Chemically, plant cell walls contain several sugar residues, including arabinose, mannose, glucose, galactose, and fructose, which are organized as polysaccharide with different structures, forming complexes with water, lignin, and protein (Wang and McAllister 2002). Analytically, plant polysaccharides consist of cellulose, hemicelluloses, and pectin. There is considerable diversity in cell wall structures, but cellulose is most often present as tightly bundled microfibrils, noncovalently bonded together by hemicelluloses; pectins are usually less prevalent (van den Brink and de Vries 2011). The presence of lignin can limit the enzymatic breakdown of cellulose and other polysaccharides, by adsorption of the enzymes and blocking access of the enzymes to the polymers (Jørgensen et al. 2007). For efficient hydrolysis of the constituent polymers of plant cell walls, these enzymes must overcome the structural barriers imposed by the cell walls (Gupta et al. 2016) and stimulate the release of free oligosaccharide and monosaccharides (Da Silva et al. 2017).

Fungi secrete a repertoire of enzymes, which they use in degrading organic matter, such as trees, leaves, and other vegetation under natural conditions, and can be exploited by man to degrade fibrous materials, including sawdust, agricultural and forestry residues, and agroindustry by-products. Considering the existing intermolecular cross-links, cell wall digestion requires that hydrolytic enzymes can cleave the bonds within the cross-linked matrix (Wang and McAllister 2002; Rodrigues et al. 2008). Most fungal enzymes have a carbohydrate-binding module, which exhibits binding preference for cellulose, thus increasing the association of the enzyme with amorphous and crystalline celluloses (Tomme et al. 1995; Suurnäkki et al. 2000).

The classification of hemicelluloses is based on the main sugar residue in the backbone of the polymer, such as mannan (β-1,4-linked D-mannose), xylan (β-1,4-linked D-xylose) and xyloglucan (β-1,4-linked D-glucose) (van den Brink and de Vries 2011). Hemicelluloses are intimately associated with lignin, which makes fiber degradation difficult (Kung 2006). The enzymatic breakdown of hemicellulose requires various set of enzymes acting on both terminal and/or internal glycosidic linkages (Gupta et al. 2016). They include non-arabinose- and arabinose-liberating endo-xylanases, β-xylosidases, acetyl xylan esterases, endo-xylanases, α-glucuronidases, β-mannanases, xyloglucanases, and α-galactosidases (Gupta et al. 2016). In addition, the breakdown of hemicellulose also depends on microbial enzymes due to its composite structure such as xylose linked by -1,4 bonds (Graminha et al. 2008).

Xylanases (EC 3.2.1.8) are frequently used in feeds to damage the xylan structure of cell wall to release short-chain sugars (Østergaard and Olsen 2010). Xylan is the primary compound of the hemicellulose component of hardwood after cellulose (Rahman et al. 2003; Martínez et al. 2005). It consists of a homopolymeric backbone of 1,4-linked β-D-xylopyranose units, and depending on its origin, the backbone may be substituted with different groups, such as glucuronosyl, acetyl, and arabinosyl residues; arabinosyl residues are sometimes replaced by coumaryl and ferulolyl residues, crosslinking hemicellulose to lignin (Rahman et al. 2003). Xylanases are typically endo-type enzymes and attack polymer chains in a random manner, causing a decrease in the level of polymerization of the substrate. They work in conjunction with other debranching enzymes, such as phenolic acid esterase, acetyl esterase, α-glucuronosidase, and α-L-arabinofuran-osidase, which eliminate the branching chain substituents of heteroxylans, to produce the polymers' constituent sugars (Rahman et al. 2003). Fungal enzymes perform better under somewhat acidic conditions (Vikramathithan et al. 2012).

Aspergilli produce a wide range of enzymes capable of degrading pectin (van den Brink and de Vries 2011). Pectin is not as predominant as cellulose and hemicellulose in plant biomass, and they can be found in the cell walls of plants and cementing material for cellulose networks in apples (Dhillon et al. 2011) and is abundant in citrus peel and pulp. The structure of pectin consists mainly of α-1,4-linked D-galacturonic acid residues that can be substituted with acetyl groups or methyl-esterified group (van den Brink and de Vries 2011). The structural degradation of pectin requires two groups of enzymes: glycoside hydrolases and polysaccharide lyases. Fungal glycoside hydrolases involved in the degradation of the pectin backbone belong to the GH family 28 (Martens-Uzunova and Schaap 2009). These GH28 enzymes, endo- and exo-polygalacturonases, attach to the smooth regions of the pectin backbone, while the more complex, stubbly regions of the pectin are further degraded by xylogalacturonase, unsaturated glucuronylhydrolases, α-rhamnosidases, endo- and exo-rhamnogalacturonases, and unsaturated rhamnogalacturonanhydrolases (van den Brink and de Vries 2011). In general, endo- and exo-polygalacturonases split the α-1,4-glycosidic linkages between the α-galacturonic acids (van den Brink and de Vries 2011). Furthermore, pectin esterase is the enzyme that hydrolyzes the ester links between carboxyl and methyl groups of pectin (Ceci and Lozano 1998).

Trichoderma reesei is well known as being a cellulose-degrading fungal species, whereas *Phanerochaete chrysosporium* is a fungus effective at degrading lignin (van den Brink and de Vries 2011). Of plant cell wall polymers, cellulose is the least complex, with a linear structure of β-1,4-linked D-glucose residues (van den Brink and de Vries 2011). Enzymatic cellulose hydrolysis involves cellobiohydrolase enzymes, which break the cellulose linkages from the ends to produce cellobiose, and endo-β-1,4-glucanases, which cut the cellulose polymer randomly, thereby reducing the polyglucan chain and generating reactive ends for cellobiohydrolases (Peciulyte et al. 2017). The cellulase activity of *T. reesei* is composed of endoglucanases and exoglucanases that act in synergy to catalyze cellulose to cellobiose (Okada et al. 1998). Bioconversion of lignocellulose waste to soluble sugars depends on the synergistic actions of complete cellulase systems involving carboxymethyl cellulase, cellobiohydrolase, and cellobiose (Ikram-ul-Haq et al. 2006). The hydrolysis of cellulose is distinct from most other enzymatic reactions, because it involves soluble enzymes acting on an insoluble substrate with a complex supramolecular structure (Peciulyte et al. 2017). Furthermore, cellulose degradation involves substrate-adsorbing cellobiohydrolase and endoglucanase. These adsorbed enzymes release the cellulase molecule from the cellulose chain and release cellobiose and cellodextrin from the cellulose molecule; the cellobiose and cellodextrin are converted into glucose by β-glucosidases (Bansal et al. 2009). Exoglucanases and β-glucosidases are severely inhibited by cellobiose and glucose, respectively (Gupta et al. 2016).

Ferulic acid (4-hydroxy-3-methoxycinnamic acid) is the primary phenolic acid existing in abundance in plant cell walls, where it crosslinks molecules of arabinoxylan through arabinose residues (Uraji et al. 2014). Ferulic acid production from cereal bran is an example of the use of fungi to obtain a high-valued commodity from agricultural by-products (Long et al. 2016). Feruloyl esterases (EC 3.1.1.73) can break the ester linkages between ferulic acid and polysaccharides (Topakas et al. 2007; Gong et al. 2013), and it has been acknowledged to be a key enzyme in releasing ferulic acid from plant materials (Uraji et al. 2014). In another study (Mohanram et al. 2015), pretreatment of rice straw by *T. hirsuta* and *M. roridum* yielded 1.7 g/kg and 2.0 g/kg of phenolic compounds, respectively. Although the work of Mohanram et al. (2015) examined the pretreatment of rice straw as a source of biomass for biofuel production, their results suggested that *T. hirsuta* and *M. roridum* have potential application in the pretreatment of feeds and forages to improve nutritive value by increasing the availability of digestible polysaccharides and sugars, as well as reducing levels of antinutritional factors. This could be important in countries where rice production is prominent and their local ruminant feed resources contain antinutritional constituents.

Lignin in the layers of a plant cell wall, alongside hemicellulose and cellulose fibrils, forms an amorphous matrix, which severely limits its biodegradation (Fengel and Wegener 1984). This accounts for the high mechanical strength of wood and its extraordinary resistance to biological and chemical degradation (Martínez et al. 2005). Lignocellulose-degrading processes are initiated when extracellular lignin-modifying enzymes rupture the intact lignin (Gupta et al. 2016). Lignin-degrading microbes produce lignases and a few other enzymes, such as laccases and peroxidases, which work in synergy just to break down lignin into its phenolic building blocks (Gupta et al. 2016). These enzymes target lignin carbon–oxygen and carbon-carbon bonds and β-1 linkages, which hold together the complex three-dimensional lignin structure (Berlin and Balakshin 2014). To facilitate the release of plant cell wall carbohydrates from lignified plant biomass for industrial purposes, the material can be pretreated to break down the lignin-carbohydrate complexes, consequently making the matrix more vulnerable to enzymatic hydrolysis (Mosier et al. 2005; Wan and Li 2010). Figure 1 summarizes the breakdown process of lignocellulose by fungal enzymes.

Alpha amylase and glucoamylase are the predominant enzymes that degrade starch. Glucoamylase (Exo-1,4-α-D-glucan glucanohydrolase) hydrolyzes single glucose residues from the nonreducing ends of amylose and amylopectin in a stepwise manner. It randomly hydrolyzes α-1,4-glucosidic bonds in amylose and amylopectin molecules of starch. Since glucoamylase cannot cleave the α-1,6-linkages in amylopectin molecules, a diverse range of products, including dextrins and smaller polymers composed of glucose units, are formed (Windish and Mhatre 1965).

Esterases are enzymes capable of breaking ester linkages (i.e., linkages form between hydroxyl and carboxyl groups). Peptidase enzymes break peptide bonds within proteins into peptides and amino acids. Both enzyme types are used by all organisms in the degradation process (Da Silva et al. 2017).

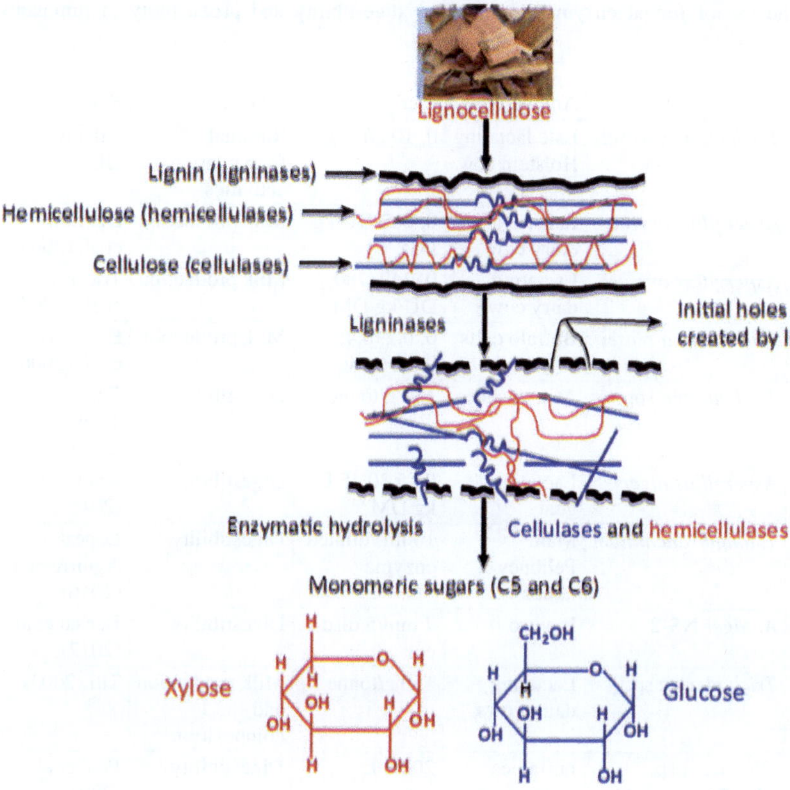

Fig. 1 The breakdown of lignocellulose by fungal enzymes (Gupta et al. 2016)

3 Exogenous Fungal Enzymes and Feed Utilization

3.1 Feed Intake

Voluntary feed intake is one of the most important factors influencing livestock productivity. Enzyme supplementation of feeds of lactating buffalos, cows and male calves was shown to result in an increase in dry metter (DM) intake. (Tables 1 and 2 Azam et al. 2017). Using Pelibuey lambs, Lopez-Aguirre et al. (2016) reported that supplementing a total mixed ration (based on sorghum grain and buffel grass) with an enzyme preparation from *T. longibrachiatum* (formerly *Trichoderma reesei*) containing cellulase and β-glucanase had no influence on DM intake (Tables 1 and 2).

Table 1 The use of fungal enzymes to influence digestibility and productivity in ruminants (Part 1)

Enzyme	Fungus source	Animal type	Rate or levels used	Influence	References
Amylase Xylanase	*Trichoderma reesei*	Late lactating Holstein cow	0, 10 g/h/day	Ruminal fermentation activities	Hristov et al. (2008)
Amylase	*Aspergillus oryzae*	Lactating dairy cow	0, 662 DU/kg DM	Milk production	DeFrain et al. (2005)
Amylase	*Aspergillus oryzae*	Lactating dairy cows	0, 240, 780, DU/kg DM	Milk production	Tricarico et al. (2005)
Amylase Xylanase	*Trichoderma reesei*	Buffalo calfs	0, 0.2, 0.4, 0.6% w/w	Milk production	El-Kady et al. (2006)
Cellulase	*Trichoderma* spp.	Awassi sheep	150 g/tonne	Digestibility	Titi and Tabbaa (2004)
Cellulase	*Aspergillus niger*	Lactating goat	0,15,30,45 U/kg DM	Digestibility	Kholif et al. (2015)
Cellulase	*T. longibranchiatum*	Male Pelibuey lambs	10 mL diluted enzyme	Digestibility	Lopez-Aguirre et al. (2016)
Cellulase	*A. niger* NS-2	In vitro	7 mm/5 disc cut	Digestibility	Bansal et al. (2012)
Cellulase	*Trichoderma* sp.	Lactating dairy cows	150 g/tonne	Milk production and composition	Titi (2003)
Fungal culture	*Piromyces* sp. FNG5	buffaloes	200 mL	Digestibility	Paul et al. (2004)
Laccase	*Pleurotus ostreatus*, *Trametes versicolor*	Inoculation of feed/in vitro	0.5% w/w	Digestibility, ruminal fermentation activities	Shrivastava et al. (2011)
Not specified	*A. niger*	Inoculation/ in vitro	0.3% w/w	Digestibility	Wang et al. (2006)
Not specified	Fungal strain ABO 374	Merino mutton lambs	7.5 mL/kg	Growth performance	Cruywagen and van Zyl (2008)

3.2 Digestibility

Forage digestibility is a major factor limiting the nutritive value and voluntary intake of ruminant diets. Plant cell walls comprise 40–70% of the DM of typical forages, and even under ideal feeding conditions, cell wall digestibility in the total digestive tract is still generally less than 65% (Van Soest 1994). Furthermore, excreta from poorly digested feeds and other agricultural and agroindustry by-products contribute to environmental waste, and their degradation results in increased emission of greenhouse gases, carbon dioxide, and methane.

Table 2 The use of fungal enzymes to influence digestibility and productivity in ruminants (Part 2)

Enzyme	Fungi source	Animal type	Rate or levels used	Influence	References
Protease Pectinase Xylanase	*Aspergillus awamori*	buffaloes	40 g/h/d	Digestibility/ milk composition	Morsy et al. (2016)
Xylanase	*Trichoderma reesei*	Mid-lactating cows	0, 8, 16, 24 g/h/d	Ruminal fermentation activities	Silva et al. (2016)
Xylanase	*Aspergullus niger/T. longibranchiatum*	Lambs	0, 10 g/d	Ruminal fermentation activities	Ebtehag et al. (2016)
Xylanase	*Aspergillus niger* BC-1, *T. reesei* rut C-30	Inoculation	1×10^7 spores/gram dry substrate	Digestibility, ruminal fermentation activities	Dhillon et al. (2011)
Xylanase	*T. longibranchiatum*	Frisian dairy cows	0, 2.15, 4.30 kg/ dry matter (DM)	Ruminal fermentation activities	Miller et al. (2008)
Xylanase	*Trichoderma reesei*	Sheep	3.9 mL/kg total mixed ration	No effect	Peters et al. (2015)
		Lactating Holstein cows	3.8 mL/kg total mixed ration	No effect	
Xylanase Endoglucanase Exogucanase	*T. longibranchiatum*, *P. funiculosum*	In vitro	0, 0.5, 1, 2, 4 UI/g	Digestibility	Phakachoed et al. (2013)
Xylanase Cellulase	*Aspergillus niger* *Trichoderma viride*	Nili Ravi buffalos	0,10, 15, 20 g/h/d	Milk production	Azam et al. (2017)
Xylanase Cellulase	Not specified	Lambs	0.35% v/w	Digestibility	McAllister et al. (2000)

The primary purpose of adding enzymes to ruminant feeds is to improve digestibility of feeds either by pretreatment of the feed or by addition prior to feeding to increase breakdown within the digestive tract. Ideally, an increase in digestibility would lead to improved production efficiency, better feed conversion rate, and lowered cost of production. While bacteria, archaea, protozoa, and fungi are responsible for breakdown of plant material in the rumen, exogenous enzymes of fungal origin may be useful for the catalysis of fibrous material. Several studies have shown that exogenous fungal enzymes increased diet digestibility, evidenced by increased digestibilities of DM, organic matter (OM), crude fibre (CF), neutral detergent fiber (NDF), and acid detergent fiber (ADF) - (Mertens and Huhtanen 2007; Morsy et al. 2016).

Titi and Tabbaa (2004) used a commercial enzyme preparation containing 200 unit/mg cellulase produced by *Trichoderma* spp. to treat feeds given to Awassi sheep. This resulted in increased digestibility of DM (66 vs. 64%), OM (78 vs. 75%), CF (47 vs. 45%), NDF (43 vs. 40%), ADF (45 vs. 41%), and N-retained (10 vs. 9 g/animal/day) in the treated versus the untreated control. The increase in digestibility was likely due to fibrolytic activity of *Trichoderma* spp., while the higher N-retention perhaps resulted from increased microbial protein synthesis. Furthermore, the study on male dairy calves showed that total gain increased ($p < 0.03$) by 20% and feed conversion ratio decreased ($p < 0.04$) by 9.41% in calves supplemented with cellulase enzyme.

Morsy et al. (2016) reported improved digestibilities of DM, Crude protein (CP), ether extract (EE), nonstructural carbohydrates, NDF, and ADF after adding *A. awamori*-based enzyme product to the diet (70% forage) of buffaloes. Kholif et al. (2015) examined the effect of adding a commercial cellulase enzyme (Asperozym) produced from *A. niger* to the diet of lactating goats. The cellulolytic enzyme increased nutrient digestibility (OM, ADF, and acid detergent lignin (ADL)) and the nutritive value (total digestible nutrients: 67% for the enzyme-supplemented diet vs. 56% for the control diet).

Paul et al. (2004) fed an anaerobic fungus (*Piromyces* strain FNG5) isolated from the faeces of wild blue bull (*Boselaphus tragocamelus*) to buffaloes (*Bubalus bubalis*). Results showed that the fungus fed animals had increased DM, OM, NDF, and ADF digestibility by 7.77%, 7.17%, 15.70%, and 17.68%, respectively. Fungal enzymes have a wide range of actions, which are manifested by their ability to simultaneously increase the digestion of protein, starch, pectin, and cellulose in the diet. Four strains of *A. niger* were tested by Wang et al. (2006) on their ability to degrade bran mash. F3 and F4 strains produced more cellulase, hemicellulase, glucoamylase, pectinase, and acidic proteinase than others. Strain ratio, moisture content of raw material, and availability of protein sources influence the enzymatic activity of *A. niger*. Fungi are well known to grow better in a wet environment (Krishna 2005). The ability of these *A. niger* strains to produce such enzymes could be taken advantage of in ruminant nutrition. The use of *A. niger* in ruminant diet could be a viable option and their efficiency can be further enhanced in the rumen by the availability of nitrogen in the form of ammonia nitrogen. It is likely that the presence of nitrogen provides a viable environment for other fibrolytic microbes; thus, as fungi degrade lignin and cellulose complex, other microbes are also releasing nutrients such as hemicellulose, pectin, through degradation (Krishna 2005; Wang et al. 2006). This mechanism of action is likely to be complementary.

Shrivastava et al. (2011) evaluated the effects of *Pleurotus ostreatus* and *Trametes versicolor* white-rot fungi on the degradation of wheat straw. After 30 days of incubation with *P. ostreatus*, there was 30.9%, 35.00%, 38.88%, 45.05%, 37.48%, and 37.86% disappearance in dry matter, acid detergent fiber, neutral detergent fiber, hemicellulose, lignin, and cellulose, respectively. Similarly, incubation with *T. versicolor* resulted in a decrease in dry matter, acid detergent fiber, neutral detergent fiber, hemicellulose, lignin, and cellulose by 24.38%, 30.04%, 33.85%, 39.89%, 31.29%, and 34.00%, respectively. In contrast, both *P. ostreatus* and T. *versicolor* increased CP by 50.74% and 39.17%, respectively. The variation in the influence of both fungi could mean that *P. ostreatus* can produce greater amounts of or more active hydrolytic enzymes (hemicellulase, lignase, and cellulase) than *T. versicolor.* Interestingly, the protein level increased after treatment with both fungi. This could be due to the availability of more nutrients from the degradation of hemicellulose, NDF, ADF, which supported proliferation of microbes. The improvement of protein could likely because the enzyme targeted nitrogen cross-linkages in the cell wall of forages (Roque et al. 2017). This is really a good discovery for tropical countries and rural farmers who are constantly challenged with the daunting task of providing quality protein for their animals due to limited availability. This could really help to prevent the seasonal forage growth pattern which is a common occurrence in the tropics and other developing nations where low protein agricultural by-product constitute the bulk of ruminants' diets. Similarly, Zhao et al. (2015) reported that fungi pretreamnent of rape straw with *P. chrysosporium* and *Lentinula edodes* for 30 days resulted in fermentation loss of about 40.8%, 41.6%, 52.9%, 52.8%, and 45.7% of DM, OM, NDF, ADF and ADL, respectively. Chemical composition of the rape straw after incubation by the treatment decreased by about 17.56% and 17.02% for NDF and ADF, respectively. In, contrast, the CP protein of *P. chrysosporium* and *L. edodes* incubated straw increased by 22.5% and 30.57%, respectively. Furthermore, El-Kady et al. (2006) using enzyme produced by *Trichoderma reesei* F- 418, which contains (cellulase 114.2, xylanase 662.1, alpha-amylase 1065.1, and polyglacturenase (pectinase) 216.2 IU/g fermented sugar beet pulp). The sugar beet pulp was applied at 0.2%, 0.4%, and 0.6% (W/W) to a concentrate feed mixture and fed to buffalo male calves. Nutrient digestion (CP, EE, nitrogen free extract (NFE), and total digestible nutrients (TDN)) of the enzyme group was better than the control, although digestion was highest at 0.6%. Nutrient digestion (DM, OM, and CF) was higher at 0.6% supplementation than control and other supplementation level. Miller et al. (2008) investigated the effects of adding at low or high levels, a commercial enzyme product (Roxazyme® G2 Liquid) on milk production and rumen fermentation characteristics in grazing dairy cows fed barley or sorghum grain-based supplements. The enzyme product consisted of a mixture of xylanase and endoglucanase exogenous enzymes derived from *T. longibrachiatum.* The enzyme added to the barley supplement resulted in higher ruminal total volatile fatty acids (VFA), acetate, and butyrate concentrations, suggesting increased digestibility. In contrast, higher level of enzyme addition to the sorghum grain resulted in reductions in the ruminal concentrations of total and individual (acetate, propionate, and butyrate) VFA. A pre-treatment of sorghum with enzyme could possibly reduce the levels of its anti-nutritional factors.

4 Chemical Pretreatment and/or Treatment

Efficient conversion of lignocellulosic biomass to fermentable carbohydrates is difficult to achieve without some form of pretreatment (Narra et al. 2012). Bansal et al. (2012) studied the ability of *A. niger* to produce cellulase from lignocellulosic substrates (e.g., potato peelings, rice husk, sawdust, corn cobs, carrot peelings, kitchen waste, grasses, leaves, pineapple waste, orange peelings, wheat bran, wheat straw) pretreated with water, or solutions of NaOH or H_2SO_4. Higher cellulase activity was exhibited in NaOH-treated substrate than that treated with H_2SO_4 or water. Higher cellulase production is an indication of increased cellulose digestibility, which is possible due to the breaking of linkages between the lignocellulose complexes (Rahman et al. 2003). Wang et al. (2006) treated wheat bran with KH_2PO_4 and $(NH_4)_2SO_4$ and found an increased enzymatic activity of cellulase, hemicellulase, glucoamylase, pectinase, and acidic proteinase at 0.1%, 0.3%, 0.5%, 1%, and 1.5%, respectively. Pretreatment/treatment with chemicals enhances extraction of useful nutrient from poor-quality feeds.

Hassan et al. (2012) evaluated the influence of urea and fungi (*P. ostreatus*) on barley straw. In vivo digestibility showed that barley straws treated with urea and fungi had increased OM, CF, ADF, and cellulose by 5.48%, 4.69%, 6.56%, and 8.2%, respectively. Higher digestibility both in vivo and in vitro may be attributed to the ability of fungi to break down the cell wall content, which could have increased surface area for further microbial attachment and breakdown. Trach et al. (2001) treated rice with 3% quicklime, 4% urea, and their combination, in some tropical young bulls. Chemical treatment improved straw OM digestibility by 10.19%, 7.69%, and 14% for urea, quicklime, and their combination, respectively. Furthermore, crossbred dairy heifer of Friesian and zebu had improved straw digestibility of 11.89% and 15.39% when fed with 4% urea and a 3% quicklime +2% urea treated straw, respectively. The variations in response to both treatments may be because N supplied by urea-treated straw was more efficiently incorporated into bacterial protein than N in the form of urea sprayed just prior to feeding. In Gunun et al. (2013), rice straw pretreated with urea increased ($P < 0.05$) apparent digestibility of DM, OM, NDF, and ADF by 10.89%, 9.50%, 12.61%, and 12.44%, respectively, while CP increased by 34.9% in dairy steers. Ammonium hydroxide (NH_4OH) formed in urea-treated rice straw produces a swelling of the hemicelluloses-lignin complex in rice straw (Mapato et al. 2010). Hence, the swelling leads to increased surface area for penetration of microbes, which increased digestion.

Thi Huyen et al. (2019) pretreated rice straw with a fungus *Pleurotus eryngii* (*P. eryngii*; strain MES 03757) incubated for 28 days had 17.4%, 33.9%, 28.9%, and 47.7% reduction in OM, NDF, ADF and acid detergent lignin respectively compared to untreated rice straw whereas crude protein increased by 70.6%. In vivo digestibility with sheep showed that when 30% of pretreated rice straw was added to the forage and fed to sheep, DM, CP, NDF, and ADF apparent digestibility increased by 5.51%, 6.64%, 4.5%, and 4.24%, respectively. The increase in CP may be due to the concentration effect of biomass, proliferation of fungi, or other microbes, which were added to the nutrient. Similarly, Vorlaphim et al. (2018) pretreated rice stubble with three white-rot fungi of *P. ostreatus*, *P. sajor-caju*, and *P. eous*. Chemical composition after incubation showed that all treatments were statistically similarly to one another but different from the control for CP, NDF, and ADF contents, which averagely increased by 32.8% and decreased by 15.79% and 12.87%, respectively. However, the ADL of all treatments was higher than the control but different from one another. Treatment with *P. ostreatus*, *P. sajor-caju*, and *P. eous* resulted in 18.78%, 41.63%, and 33% decrease in ADL. Further in vitro gas study showed that gas production of rice stubble treated with *P. ostreatus*, *P. sajor-caju*, and *P. eous* increased by 305%, 138.86%, and 187.34%, respectively, compared to rice stubble without pretreatment. Also, OMD was reduced by 25.58%, 29.04%, and 18.07% for *P. ostreatus*, *P. sajor-caju*, and *P. eous*, respectively. The increase in gas production may be attributed to the ability of ruminal microbes to access the organic matter embedded within the cell wall, which had been broken down by the fungi. Thus, it may be accepted that one of the main benefits of pretreating straws and stubble with fungi is improved crude protein and digestibility. This may be beneficial in areas with poor access to quality feed. This could also encourage interest in using inedible, recalcitrant fodders from farms or encourage recycling of nutrients to feed ruminant among smallholder farmers.

5 Ruminal Volatile Fatty Acids

Studies have shown that fungal treatment of fibrous feed materials results in increased feed digestibility. In growing lambs, Ebtehag et al. (2016) investigated the effects of supplementing a mixed concentrate/forage diet with a commercial enzyme product obtained from the fermentation extract of *A. niger* and *T. longibrachiatum*. Enzyme supplementation increased total volatile fatty acid (VFA) concentrations in the rumen.

Silva et al. (2016) evaluated the effects, in mid-lactating cows receiving a total mixed ration (based in maize silage and ground maize), of administering in the morning feed, either 0, 8, 16, and 24 g/head/d, of a commercial enzyme derived from *T. longibarachiatum*. The rumen fluid was collected at multiple sites (anterior dorsal, anterior ventral, medium ventral, posterior dorsal, and posterior ventral) within the rumen before feeding and 2, 4, 6, 8, 10, 12, 14, and 16 h post feeding. There was a progressive increase in the production of ruminal short chain fatty acids (acetate, propionate, and butyrate), with enzyme addition rate.

The alteration of ruminal VFA fermentation patterns by dietary supplementation with fungal enzymes is apparently not limited to those enzymes capable of breaking down lignocellulose) Tricarico et al. (2005) observed increases in the acetate: propionate molar concentration ratio in the rumen of steers and lactating dairy cattle supplemented with exogenous fungal α-amylase. In addition, rice straw pretreated with three strains of white rot fungi (*P. tuberregium*, *P. pulmonarius*, and *P. ostreatus* by (Wuanor and Ayoade 2017) resulted in 3-, 3-, and 4.33-fold increase in short-chain fatty acid (SCFA) than the untreated control. Short-chain fatty acids (SCFA) are by-products of microbial fermentation, and the observed increase with fungi treatment shows that there was increased microbial activity, which was made possible due to increased access to organic matter.

6 Ruminal NH_3-N Concentration

Among other factors that could affect rumen microbes, rumen NH_3-N concentration can be a limiting factor that affects digestion of feeds (Perdock 1987), since many rumen microbes require NH_3-N for their growth. Furthermore, rapid breakdown of dietary protein to NH_3 can increase nitrogenous wastage if not coupled with a concurrent supply of fermentable energy, to allow microbial proliferation. High rumen NH_3 concentration can reduce dry matter intake and thus, productivity of the animal, and in extreme cases, NH_3 toxicity and death may occur. Ebtehag et al. (2016) supplemented the diet of growing lambs with a commercial xylanase enzyme (Fibrozyme) compounded from the fermentation extract of *A. niger* and *T. longibrachiatum* and reported higher ruminal concentration of NH_3-N at 0, 3, and 6 h post-feeding than that of animals receiving no enzyme supplement. This suggests that the xylanase had the ability to release nitrogen embedded in the cell wall of plants.

7 Microbial Protein Synthesis

The complex stomach of ruminants allows them to convert dietary protein, which may be of poor quality, and nonprotein nitrogen into higher-quality microbial protein (Dewhurst et al. 2000). Microbial protein synthesis depends upon various factors such as sources of carbohydrate and protein, the level of voluntary feed intake, the recycling of microbes in the rumen, and the antinutrients of plants consumed (Ekanem 2016). The synthesis of microbial protein can be influenced by the availability of readily fermentable carbohydrate in the diet (Dewhurst et al. 2000). Enzyme pretreatment to enhance the availability of substrate by increasing the extraction of nutrients by breaking the cell wall of plant materials can also encourage microbial proliferation. Paul et al. (2004) fed buffaloes (*B. bubalis*) with an anaerobic fungus (*Piromyces* strain FNG5), isolated from the feces of the wild blue bull (*B. tragocamelus*). Results showed an increase in nitrogen intake (90.2 vs. 79.9 g/day), nitrogen retention (25.8 vs. 15.0 g/day), NR (288 vs. 191 g/kg N intake), and NR (399 vs. 262 g/kg N absorbed). The fungus improved the rumen microbial population (fungi, cellulolytic bacteria, hemicellulolytic bacteria, total viable bacteria) in the treated versus the untreated buffaloes. Microbial increase may be attributed to increased feed intake, increased degradability, which released more monosaccharide, disaccharide, and lignin-bound substrates. Also, availability of N retained in fungus may be the reason for their increase. Hence, enhanced N availability improved digestibility. Nitrogen retention helps in reducing the excretion of nitrogen into the environment and thus contributes to greenhouse gas mitigation.

8 Milk Production and Composition

The performance of an animal depends largely on feed intake and its ability to extract nutrients from feeds. In addition to the genetic makeup of the animals, profitable milk production depends on the ability to partition nutrients for its maintenance and production. Milk composition is influenced by the products of ruminal fermentation. For instance, VFA fermentation patterns in the rumen can affect milk fat levels and milk output. Furthermore, the fatty acid profile of milk is to some extent affected by the lipid profile of the diet and the extent to which dietary lipid is bio-hydrogenated by rumen microbes.

Exogenous enzymes can decrease the ωn-6: ωn-3 ratio of the fatty acids in milk fat. While this is desirable, increasing conjugated linoleic *acid levels* in animal products is also of great interest due to its benefit to human health. Higher consumption of ωn-3 fatty acid has been associated with benefits to psychiatric disorders (Freeman 2000) and cardiac health (Breslow 2006).

In studies with dairy cattle, Titi (2003) tested a dietary supplement of a cellulase/hemicellulase commercial enzyme preparation (Maxicel 200L®). Enzyme inclusion increased milk production, 3.5% fat-corrected milk production and energy-corrected milk production over the 100-day measurement period. Milk fat and protein levels were also raised because of enzyme supplementation. These effects could be attributed to an increase in digestibility of feed consumed by the cows, the enzyme, and increased digestibility of NDF.

Tricarico et al. (2005) used a commercial extract of *A. oryzae* with α-amylase activity added to a maize-based total mixed ration fed to Holstein cows. Milk yield of the control versus α-amylase group was 29.2 versus 30.7, 30.4, 29.6 kg/d respectively, and milk fat was 37.1 versus 37.8, 37.8, 37.5 g/kg, respectively, at ($p = 0.02$) quadratic significance level. The reason might be that the glucogenic acid proportion was within tolerance level of the fibrolytic bacteria such that those microbes use the glucose substrate available for multiplication/proliferation. It may be concluded that starch at lower level is not toxic but beneficial for growth of fibrolytic microbes. However, authors suggested that the effects were due to the increase in acetate: propionate ratio and butyrate in the rumen, which was caused by interaction of enzyme supplement with microbial metabolism. This lower enzyme application might be useful for dairy animals in early lactation when their feed intake cannot meet up to the energy demand of their body for milk production. Such enzyme may be used to prevent massive weight or quick loss of body scores due to the animal using the body energy reserve for milk production.

In Azam et al. (2017), a commercial exogenous enzyme Fibrozyme™ was given to lactating Nili Ravi buffaloes at a rate of 0, 10, 15, and 20 g/head/day. The enzyme had xylanase (43.4 IU) and cellulose (31.0) activities and was produced from *A. niger* and *Trichoderma viride* fermentation extracts. On performance, milk yield, milk protein, milk fat, solid not fat, and lactose in all the enzyme-treated cows produced better than the control. On milk composition, enzymatic supplementation increased milk protein, milk fat, solids not fat, and lactose with 15 g/h/day producing better than other levels of enzyme. Morsy et al. (2016) used a commercial enzyme Tomoko® (contain protease, pectinase, xylanase, e-amylase at 1000, 30, 25, 20 unit/g, respectively) produced by *A. awamori*, to improve total protein, casein, and true protein of milk compared to the control, which might be due to the high proteolytic activities of the enzyme in the diet. Similarly, there was a decrease in nonprotein nitrogen and urea-nitrogen, which might be due to increased microbial protein synthesis (Morsy et al. 2016), which increased as microbial proliferation increases. The increase in microbes might be the reason for the decrease in urea-N content of the milk as the urea in the rumen would have been used for their own multiplication, thereby decreasing the abundance of circulating urea in the blood. Furthermore, *A. awamori* enzyme improved lysine, total essential amino acids of the milk. In addition, monounsaturated and polyunsaturated fatty acid concentrations of the supplemented group were better than the control treatment. This improvement may be attributed to increased digestion of fiber content in the diet, because butyric acid, which is always high in fiber diets, is important in milk fat synthesis.

9 Growth Performance and Meat Quality

The ability of ruminants to convert plant materials, inedible for humans, to meat, milk, and other livestock products is of great agricultural significance (Meale et al. 2014). Enzymes can facilitate feed use by enhancing ruminal digestion, and degradation, by stimulating ruminal microbial growth (Beauchemin et al. 2004). Cruywagen and Goosen (2004) fed wheat straw pretreated with 10 mL, 5 mL, and 1 mL supernatant/kg wheat straw of a locally produced enzyme product, ABO 374, to Döhne Merino ram lambs. Body weight gain of lambs fed with wheat straw treated with 1 mL was not different from the control, while 10 and 5 mL supplementation improved by 24.77% and 31.79%, respectively. Similarly, addition of a cocktail of supernatant of the fungal strain ABO 374 (*A. terreus* var. *carneus*) containing xylanase and cellulase, either in liquid or freeze-dried forms, to high and low forage diets fed to Merino lambs resulted in higher body weight gain compared with the unsupplemented control ration (Cruywagen and van Zyl 2008). For both diets, the feed conversion ratio in the sheep on the fungal supplement was lower than that of the sheep receiving freeze-dried supplement and the control sheep. The superiority of the liquid fungal supplement compared to the freeze-dried fungal enzyme preparation may be attributed to the physical nature of enzymes. Since the dietary substrates for the enzyme preparation (hemicellulose, cellulose, and xylan) are not on the surface of feed ingredients (but are usually embedded within the cell wall), it is likely that the liquid supplement was more readily able to diffuse into the inner layer of the cell wall where it hydrolyzes the substrates. It is also possible that the freeze-drying process in preparing the dry fungal enzyme supplement could have altered the structure of the enzyme or even bind it together, thereby reducing its efficacy. When there is no increase in feed intake but there is a corresponding increase in growth rate, milk yield, and other performance parameters, it is an indication of better digestibility and more efficient nutrient utilization. Therefore, liquid form of fungi may be preferable; it is also less expensive than the solid form.

Ebtehag et al. (2016) used an enzyme compounded from the fermentation extracts of *A. niger* and *T. longibrachiatum*, given to growing lambs at 10 g/head/day. The enzyme improved the final body weight, total body weight gain, and average daily body weight gain in comparison to a control group of lambs receiving no enzyme supplement. This is an indication of better feed nutrient utilization as the enzyme did not influence the feed intake.

El-Kady et al. (2006) used enzyme produced by *T. reesei* F-418, which contain cellulase 114.2, xylanase 662.1, alpha-amylase 1065.1 and polyglacturenase (pectinase) 216.2 IU/g fermented on sugar beet pulp. The sugar beet pulp was applied at 0.2%, 0.4%, and 0.6% (W/W) to concentrate feed mixture fed to buffalo male calves. Final body weight, total body weight gain, daily body weight gain, and feed conversion ratio was better than the control. Such improvement could be attributed to a corelationship of enzyme with microbes, which improved digestion. Such relationship might have increased the rumen microbes abundantly due to the sugar beet pulp, which was a carrier for the enzyme, being a source of readily fermentable sugar source. Moreover, the high content of α-amylase in the enzyme produced must have been responsible for breaking down the simple sugar.

Titi and Tabbaa (2004) investigated the use of a cellulase enzyme preparation (Maxicel 200®) produced by *Trichoderma* spp. in diets for weaned male and female Friesian calves. The enzyme improved the final body weight (461.2 vs. 414.0 kg), total weight gain (271.9 vs. 226 kg), average body weight gain (38.4 vs. 32.3 kg), average daily weight gain (1.39 vs. 1.15 kg/head/day), feed conversion ratio (7.7 vs. 8.5) in male calves receiving cellulase versus controls, respectively. However, there were no significant effects of enzyme supplementation in female animals.

10 Other Benefits of Fungal Enzymes

Fungi are one of the primary producers of toxins in the field and during storage. These toxins are secondary fungal metabolites known as mycotoxins (Bryden 2012) Many fungi produce a range of compounds (mycotoxins) that can be toxic to livestock when ingested. Thus, fungal toxins can have negative consequences on animal production and health. The concentration of mycotoxins in animal feeds depends on whether the environmental conditions are favorable for fungal growth. Examples of toxins produced by fungi include aflatoxin, zearalenone, deoxynivalenol, fuminosins, and ochratoxin A. However, some enzymes (including those produced by fungi) and some feed additives can reduce or bind toxins to prevent their effects on animal health. A purified enzyme preparation from *P. ostreatus* culture supernatant was reported to detoxify aflatoxin B1 by opening the lactone ring (Motomura et al. 2003). Furthermore, *Armillariella tabescens*, a medicinal fungus, has multienzyme capabilities of converting and detoxifying aflatoxin. Cao et al. (2011) reported that *A. tabescens* enzyme aflatoxin-oxidase was able to degrade aflatoxin B1. Thus, there is potential for such an enzyme to be useful in degradation of aflatoxin in food, grain, and some ruminant feeds contaminated with mycotoxins. Similarly, Motomura et al. (2003) reported that they were able to purify an aflatoxin-degrading enzyme from *P. ostreatus* (white-rot fungi).

11 Conclusions

Exogenous enzyme supplementation from *Trichoderma* spp. and *Aspergillus* spp. could be added to the diet of ruminants to improve the dry matter, acid detergent fiber, neutral detergent fiber, and acid detergent lignin digestion, which may be recalcitrant to endogenous enzymatic breakdown. Pretreatment of plant biomass such as straw, stalks, etc., with *Pleurotus* spp., *P. eryngii*, *P. tuberregium*, *P. pulmonarius*, *P. ostreatus*, *P. sajor-caju* and *P. eous*, *P. chrysosporium*, and *L. edodes* can improve in vivo and in vitro digestibility. Similarly, a peculiar attribute with fungi pretreated biomass is improvement in crude protein concentration. Fungal enzyme supplementation could be a useful additive in the diet of growing lamb, kids, and calves to improve performances such as final body weight, weight gain, and feed conversion ratio.

Exogenous fungal enzyme can help in the degradation of antinutritional factors. This discovery is going to be very useful for researchers, and farmers that are encumbered by the problem of antinutritional constituent of their feeds. However, to prevent much antinutritional factors in rumen degradation, it is advisable to pretreat the feeds with enzymes to reduce its antinutritional properties. In developed countries, especially in dairy farms, the inclusion of tested fungal enzyme, especially in diets of dairy cattle, has the potential to increase the yield and fat content in milk. Improvement in milk fat level is an indication of better fiber digestibility. Enzyme from white-rot fungi and *A. tabescens* can be used to degrade mycotoxin in feed and forage.

More research is needed to explore the maximum use of crop residues and industrial by-products. This will help reduce feeding grains to animals and alleviate the feed versus food competition. The hydrolytic activity of fungi and their enzyme on agricultural by-products like, bran, banana, citrus fruit, sugar cane stalk, sugar beet pulp, sorghum straw, beans pod, groundnut haulms, cocoa pod, pigeon pea pod, maize cob, and maize shaft shows that it is possible to incorporate these ingredients into feeds.

References

Azam B, Tahir MN, Shahzad F et al (2017) Exogenous fibrolytic enzymes addition in concentrate ration of lactating Nili Ravi buffaloes: effects on milk production and diet digestibility. Pak J Zool 49:1359–1364

Bansal P, Hall M, Realff MJ et al (2009) Modeling cellulase kinetics on lignocellulosic substrates. Biotechnol Adv 27:833–848

Bansal N, Tewari R, Soni R et al (2012) Production of cellulases from *Aspergillus niger* NS-2 in solid state fermentation on agricultural and kitchen waste residues. Waste Manag 32:1341–1346

Beauchemin KA, Colombatto D, Morgavi DP (2004) A rationale for the development of feed enzyme products for ruminants. Can J Anim Sci 84:23–36

Berlin A, Balakshin M (2014) Industrial lignins, analysis, properties, and applications. In: Gupta VK et al (eds) Bioenergy research: advances and applications. Elsevier, Amsterdam, pp 315–336

Breslow JL (2006) n-3 fatty acids and cardiovascular disease. Am J Clin Nutr 83:1477S–1482S

Bryden WL (2012) Mycotoxin contamination of the feed supply chain: implication for animal productivity and feed security. Anim Feed Sci Technol 173:134–158

Cao H, Liu D, Mo X et al (2011) A fungal enzyme with the ability of aflatoxin B1 conversion: purification and ESI-MS/MS identification. Microbiol Res 166:475–483

Ceci L, Lozano JE (1998) Determination of enzymatic activities of commercial pectinase for the clarification of apple juice. Food Chem 61:237–241

Cruywagen CW, Goosen L (2004) Effect of an exogenous fibrolytic enzyme on growth rate, feed intake and feed conversion ratio in growing lambs. S Afr J Anim Sci 34:71–73

Cruywagen CW, van Zyl WH (2008) Effects of a fungal enzyme cocktail treatment of high and low forage diets on lamb growth. Anim Feed Sci Technol 145:151–158

Da Silva RR, Pedezzi R, Souto TB (2017) Exploring the bioprospecting and biotechnological potential of white-rot and anaerobic *Neocallimastigo mycota* fungi: peptidases, esterases, and lignocellulolytic enzymes. Appl Microbiol Biotechnol 101:3089–3101

DeFrain JM, Hippen AR, Kalscheur KF et al (2005) Effects of dietary alpha-amylase on metabolism and performance of transition dairy cows. J Dairy Sci 88:4405–4413

Dewhurst RJ, Davies DR, Merry RJ (2000) Microbial protein supply from the rumen. Anim Feed Sci Technol 85:1–21

Dhillon GS, Brar SK, Valero JR et al (2011) Bioproduction of hydrolytic enzymes using apple pomace waste by *A. niger*: applications in biocontrol formulations and hydrolysis of chitin/chitosan. Bioprocess Biosyst Eng 34:1017–1026

Ebtehag IM, Abou Elenin EIM, Abd El-Galil ER et al (2016) Improvement of rumen fermentation and performance of growing lambs by adding natural microbial resources. Asian J Anim Sci 10:202–212

Ekanem DE (2016) Serum biochemistry, digestibility and microbial protein synthesis of West African dwarf rams fed a total mixed ration with graded levels of water washed neem fruit. MSc dissertation. University of Ibadan, Nigeria

El-Kady RI, Awadalla IM, Mohamed MI et al (2006) Effect of exogenous enzymes on the growth performance and digestibility of growing buffalo calves. Int J Agri Biol 8:354–359

Fengel D, Wegener G (1984) Wood: chemistry, ultrastructure, reactions. De Gruyter, Berlin

Freeman MP (2000) Omega-3 fatty acids in psychiatry: a review. Ann Clin Psychiatry 12:159–165

Gong YY, Yin X, Zhang HM et al (2013) Cloning, expression of a feruloyl esterase from *Aspergillus usamii* E001 and its applicability in generating ferulic acid from wheat bran. J Ind Microbiol Biotechnol 40:1433–1441

Graminha EBN, Goncalves AZL, Pirota RDPB et al (2008) Enzyme production by solid-state fermentation: application to animal nutrition. Anim Feed Sci Technol 144:1–22

Guimarães LHS, Peixoto-Nogueira SC, Michelin M et al (2006) Screening of filamentous fungi for production of enzymes of biotechnological interest. Braz J Microbiol 37:474–480

Gunun P, Wanapat M, Anantasook N (2013) Effects of physical form and urea treatment of rice straw on rumen fermentation, microbial protein synthesis and nutrient digestibility in dairy steers. Asian Australas J Anim Sci 26:1689–1697

Gupta VK, Kubicek CP, Berrin JG et al (2016) Fungal enzymes for bio-products from sustainable and waste biomass. Trend Biochem Sci 41:633–645

Hassan SA, Sadq SM, Hassan KM (2012) Evaluation of fungal or chemical treatments for barley straw in ruminants feeding 1-chemical composition, *in vitro, in vivo* digestibility and voluntary intake. Jordan J Agric Sci 8:232–241

Hristov A, Basel C, Melgar A (2008) Effect of exogenous polysaccharide-degrading enzyme preparation on ruminal fermentation and digestibility of nutrients in dairy cows. Anim Feed Sci Technol 145:182–193

Ikram-ul-Haq, Javed MM, Khan TS (2006) An innovative approach for hyperproduction of cellulolytic and hemicellulolytic enzymes by consortium of *Aspergillus niger* MSK-7 and *Trichoderma viride* MSK-10. Afr J Biotechnol 5:609–614

Jørgensen H, Vibe-Pedersen J, Larsen J et al (2007) Liquefaction of lignocellulose at high-solids concentrations. Biotechnol Bioeng 96:862–870

Kholif AM, Eman SA, Farahat MA et al (2015) Utilization of cellulolytic enzymes to improve the nutritive value of date kernels and the investigation of the impact of adding these enzymes to lactating goat's diets on rumen fermentation and nutrients digestibility. Asian J Anim Sci 9:441–447

Krishna C (2005) Solid-state fermentation systems – an overview. Crit Rev Biotechnol 25:1–30

Kung L (2006) The role of fiber in ruminant ration formulation. University of Delaware, Delaware

Lopez-Aguirre D, Hernandez-Melendez J, Rojo R (2016) Effects of exogenous enzymes and application method on nutrient intake, digestibility and growth performance of Pelibuey Lambs. Springer plus 5:1399

Long L, Ding D, Han Z et al (2016) Thermotolerant hemicellulolytic and cellulolytic enzymes from *Eupenicillium parvum* 4-14 display high efficiency upon release of ferulic acid from wheat bran. J Appl Microbiol 121:422–434

Lowe SE, Theodorou MK, Trinci APJ (1987) Cellulases and xylanase of an anaerobic rumen fungus grown on wheat straw, wheat straw holocellulose, cellulose, and xylan. Appl Environ Microbiol 53:1216–1223

Mapato C, Wanapat M, Cherdthong A (2010) Effect of urea treatment of straw and dietary level of vegetable oil on lactating dairy cows. Trop Anim Health Prod 42:1635–1642

Martens-Uzunova ES, Schaap PJ (2009) Assessment of the pectin degrading enzyme network of *Aspergillus niger* by functional genomics. Fungal Genet Biol 46:170–179

Martínez AT, Speranza M, Ruiz-Dueñas FJ et al (2005) Biodegradation of lignocellulosics: microbial, chemical, and enzymatic aspects of the fungal attack of lignin. Int Microbiol 8:195–204

McAllister TA, Stanford K, Bae HD et al (2000) Effect of a surfactant and exogenous enzymes on digestibility of feed and on growth performance and carcass traits of lambs. Can J Anim Sci 80:35–44

Meale SJ, Beauchemin KA, Hristov AN et al (2014) Board-invited review: opportunities and challenges in using exogenous enzymes to improve ruminant production. J Anim Sci 92:427–442

Mertens DR, Huhtanen P (2007) Grass forages: dynamics of digestion in the rumen. In: New York State ruminant health-nutrition conference proceedings. Syracuse

Miller DR, Granzin BC, Elliott R et al (2008) Effects of an exogenous enzyme, Roxazyme® G2 Liquid, on milk production in pasture fed dairy cows. Anim Feed Sci Technol 145:194–208

Mohanram S, Rajan K, Carrier DJ et al (2015) Insights into biological delignification of rice straw by *Trametes hirsuta* and *Myrothecium roridum* and comparison of saccharification yields with dilute acid pretreatment. Biomass Bioenergy 76:54–60

Morsy TA, Kholif AE, Kholif SM et al (2016) Effects of two enzyme feed additives on digestion and milk production in lactating Egyptian buffaloes. Ann Anim Sci 16:209–222

Mosier N, Wyman H, Dale B et al (2005) Features of promising technologies for pretreatment of lignocellulosic biomass. Bioresour Technol 96:673–686

Motomura M, Toyomasu T, Mizuno K et al (2003) Purification and characterization of an aflatoxin degradation enzyme from *Pleurotus ostreatus*. Microbiol Res 158:237–242

Narra M, Dixit G, Divecha J et al (2012) Production of cellulases by solid state fermentation with *Aspergillus terreus* and enzymatic hydrolysis of mild alkali-treated rice straw. Bioresour Technol 121:355–361

Okada H, Tada K, Sekiya T et al (1998) Molecular characterization and heterologous expression of gene encoding a low-molecular mass endoglucanase from *Trichoderma reesei* QM9414. Appl Environ Microbiol 64:555–563

Østergaard LH, Olsen HS (2010) Industrial applications of fungal enzymes. In: Hofrichter M (ed) Mycota. Industrial applications X. Springer, Heidelberg, pp 269–289

Paul SS, Kamra DN, Sastry VRB et al (2004) Effect of administration of an anaerobic gut fungus isolated from wild blue bull (*Boselaphus tragocamelus*) to buffaloes (*Bubalus bubalis*) on *in vivo* ruminal fermentation and digestion of nutrients. Anim Feed Sci Technol 115:143–157

Peciulyte A, Pisano M, de Vries RP et al (2017) Hydrolytic potential of five fungal supernatants to enhance a commercial enzyme cocktail. Biotechnol Lett 39:1403–1411

Perdock HB (1987) Ammoniated rice straw as feed for growing cattle. PhD thesis. University of New England, Australia

Peters A, Meyer U, Dänicke S (2015) Effect of exogenous fibrolytic enzymes on performance and blood profile in early and mid-lactation Holstein cows. Anim Nutr 1:229–238

Phakachoed N, Suksombat W, Colombatto D et al (2013) Use of fibrolytic enzymes additives to enhance *in vitro* ruminal fermentation of corn silage. Livestock Sci 157:100–112

Rahman AKMS, Naoyasu S, Masahiro H et al (2003) A role of xylanase, [α]-L-arabinofur- anosidase, and xylosidase in xylan degradation. Can J Microbiol 49:58–64

Rodrigues MAM, Pinto P, Bezerra RMF et al (2008) Effect of enzyme extracts isolated from white-rot fungi on chemical composition and *in vitro* digestibility of wheat straw. Anim Feed Sci Technol 141:326–338

Roque BM, Jayasooriya AD, Appuhamy RN et al (2017) Role of exogenous enzyme supplement to improve nutrition and health of ruminants. Broadening Horiz 41:1–10

Shrivastava B, Thakur S, Khasa YP et al (2011) White-rot fungal conversion of wheat straw to energy rich cattle feed. Biodegradation 22:823–831

Silva TH, Takiya CS, Vendramini THA (2016) Effects of dietary fibrolytic enzymes on chewing time, ruminal fermentation, and performance of mid-lactating dairy cows. Anim Feed Sci Technol 221:35–43

Simons PC, Versteegh HA, Jongbloed AW et al (1990) Improvement of phosphorus availability by microbial phytase in broilers and pigs. Br J Nutr 64:525–540

Sujani S, Seresinhe RT (2015) Exogenous enzymes in ruminant nutrition: a review. Asian J Anim Sci 9:85–99

Suurnäkki A, Tenkanen M, Siika-Aho M et al (2000) *Trichoderma reesei* cellulases and their core domains in the hydrolysis and modification of chemical pulp. Cellulose 7:189–109

Thi Huyen N, Tuyet Le NT, Tuan BQ (2019) Fermenting rice straw with the fungus *Pleurotus eryngii* increased the content of crude protein and the digestibility of the straw. Livest Res Rural Dev 31:25. Retrieved February 19, 2019, from http://www.lrrd.org/lrrd31/2/nthuy31025.html

Titi HH (2003) Evaluation of feeding a fibrolytic enzyme to lactating dairy cows on their lactational performance during early lactation. Asian Aust J Anim Sci 16:677–684

Titi HH, Tabbaa MJ (2004) Efficacy of exogenous cellulase on digestibility and growth of dairy calves. Livest Prod Sci 87:207–214

Tomme P, Warren RAJ, Miller RC Jr et al (1995) Cellulose-binding domains: classification and properties. In: Saddler JN, Penner M (eds) Enzymatic degradation of insoluble polysaccharides. American Chemical Society, Washington, DC, pp 142–163

Topakas E, Vafiadi C, Christakopoulos P (2007) Microbial production, characterization, and applications of feruloyl esterases. Process Biochem 42:497–509

Trach NX, Mo M, Dan CX (2001) Effects of treatment of rice straw with lime and/or urea on responses of growing cattle. Livest Res Rural Dev 13:47. Retrieved February 19, 2019, from http://www.lrrd.org/lrrd13/5/trach135.htm

Tricarico JM, Johnston JD, Dawson KA et al (2005) The effects of an *Aspergillus oryzae* extract containing alpha-amylase activity on ruminal fermentation and milk production in lactating Holstein cows. Anim Sci 81:365–374

Uraji M, Arima J, Inoue Y et al (2014) Application of two newly identified and derivatives from *Eupenicillium parvum*. J Nat Prod 71:1915–1918

van den Brink JVD, de Vries RP (2011) Fungal enzyme sets for plant polysaccharide degradation. Appl Microbiol Biotechnol 91:1477–1492

Van Soest PJ (1994) Nutritional ecology of the ruminant. Cornell University Press, Ithaca

Veum TL, Bollinger DW, Buff CE et al (2006) A genetically engineered *Escherichia coli* phytase improves nutrient utilization, growth performance, and bone strength of young swine fed diets deficient in available phosphorus. J Anim Sci 84:1147–1158

Vikramathithan J, Muthuraman P, Ravikumar S et al (2012) Biochemical and biophysical characterization of purified thermophilic xylanase isoforms in *Cereus pterogonus* plant spp. Protein J 31:141–149

Vorlaphim T, Yaungklang C, Paengkoum S et al (2018) Effect of *Pleurotus* spp pretreatment on the biodegradation of rice stubble. Livest Res Rural Dev 30:Article 204. Retrieved February 19, 2019, from http://www.lrrd.org/lrrd30/12/thans30204.html

Wan C, Li Y (2010) Microbial pretreatment of corn stover with *Ceriporiopsis subvermispora* for enzymatic hydrolysis and ethanol production. Bioresour Technol 101:6398–6403

Wang Y, McAllister TA (2002) Rumen microbes, enzymes, and feed digestion – a review. Asian Aust J Anim Sci 15:1659–1676

Wang XJ, Bai JG, Liang YX (2006) Optimization of multienzyme production by two mixed strains in solid-state fermentation. Appl Microbiol Biotechnol 73:533–540

Windish WW, Mhatre NS (1965) Microbial amylases. In: Advances in applied microbiology. Academic press, New York, pp 273–304

Wuanor AA, Ayoade JA (2017) Bioconversion of rice straw as a ruminant feed using three strains of white rot fungi. J Exp Agric Int 19:1–7

Zhao X, Gong J, Zhou S et al (2015) Effect of fungal treatment of rape straw on chemical composition and in vitro rumen fermentation characteristics. Bioresources 10:622–637

Dietary Inclusion of Exogenous Fibrolytic Enzyme to Enhance Fibrous Feed Utilization by Goats and Cattle in Southern China

Shaoxun Tang, Zhiliang Tan, and Zhixiong He

Abstract In this chapter, different exogenous fibrolytic enzymes were screened for their effect on fiber degradation, methane production, and rumen fermentation using in vitro fermentation techniques, and the selected candidates were further assessed in an in vivo study using goats and cattle feeding experiments. The results of in vitro experiments showed that forage supplemented with exogenous cellulase improved dry matter digestibility, neutral detergent fiber digestibility and total volatile fatty acid production, and decreased pH value, ammonia nitrogen, and CH_4 production. However, their optimum dosage rate varied hardly depending upon the forage. Applying exogenous xylanase to low-quality forages potentially improved rumen degradability and utilization. Furthermore, exogenous xylanase from different sources differed in their effects when applied at the same dose rate, with the responses being forage-specific. While during the in vivo studies, addition of exogenous cellulase and xylanase in goats diet did not affect the enteric methane emission and ruminal methanogen community diversity. The selected candidate exogenous cellulase inclusion in the concentrate diet of cattle did not lead to methane abatement and nutrient utilization enhancement. Thus, a more comprehensive understanding of the mode of action or mechanism of exogenous enzyme and their interaction with the other factors are needed. Further in vivo estimation is necessary after the developed enzyme products are screened with in vitro technique.

S. Tang · Z. Tan (✉) · Z. He
Key Laboratory for Agro-Ecological Processes in Subtropical Region, Hunan Research Center of Livestock & Poultry Sciences, South-Central Experimental Station of Animal Nutrition and Feed Science in the Ministry of Agriculture, Hunan Co-Innovation Center of Animal Production Safety, CICAPS, Institute of Subtropical Agriculture, Chinese Academy of Sciences, Changsha, Hunan, People's Republic of China

College of Advanced Agricultural Sciences, University of Chinese Academy of Sciences, Beijing, China
e-mail: sxtang@isa.ac.cn; zltan@isa.ac.cn; zxhe@isa.ac.cn

A. Z. M. Salem et al. (eds.), *Exogenous Enzymes as Feed Additives in Ruminants*, https://doi.org/10.1007/978-3-031-27993-5_9

Abbreviations

α-AMY	α-Amylase
A:P	Ratio of acetic acid to propionic acid
ADF	Acid detergent fiber
ALT	Alanine aminotransferase
Ca	Calcium
CEL	Cellulose
CP	Crude protein
DM	Dry matter
DMD	Dry matter degradability
DN	Digested nitrogen
DR	Dose rate
EC	Exogenous cellulase
EPX	Enzyme product with high xylanase activity
FRD0	The initial fractional rate of degradation
GE	Gross energy
GLU	Glucose
GM	Guimu-1
GP	Gas production
IVDMD	In vitro dry matter degradability
IVNDFD	In vitro neutral detergent fiber degradability
LDH	Lactic dehydrogenase
ME	Metabolizable energy
MS	Maize stover
N	Nitrogen
NDF	Neutral detergent fiber
NDFD	Neutral detergent fiber degradability
NH3-N	ammonia nitrogen
NP	*Neocallimastix patriciarum*
NS	Not significant
OM	Organic matter
P	Phosphorus
RN	Retained nitrogen
RS	Rice straw
SEM	Standard error of means
TC	Serum total cholesterol,
TG	Triglyceride
TL	*Trichoderma longibrachiatum*
TP	Total protein
TR	*Trichoderma reesei*
TVFA	Total volatile fatty acid
VFA	Volatile fatty acid
WS	Wheat straw
XYL	Xylanase

1 Introduction

Improving feed utilization, animal production performance, and decreasing nutrient discharge and greenhouse gas emissions are the persistent pursuit of animal nutrition scientists. The use of exogenous fibrolytic enzymes is one of the emerging methods that potentially improves the utilization of forages and decreases nutrient discharge by ruminants. Research has confirmed that the supplementation of enzymes to feeds can improve the degradability of dry matter (DM) and cell wall (Yang et al. 2000; Colombatto et al. 2003; Eun and Beauchemin 2007; Eun et al. 2007; Tang et al. 2008) and, consequently, the utilization of feed and production performance of ruminants (Yang et al. 2000; Arriola et al. 2011).

Although the addition of exogenous cell wall degradation enzymes to ruminant diets has been studied in an attempt to improve ruminant production, the results of the enzyme effectiveness were highly variable among various enzyme products (Beauchemin et al. 2001; Colombatto et al. 2003), even ineffectiveness (Burroughs et al. 1960; Reddish and Kung 2007) or negative responses (Sutton et al. 2003) have been frequently observed in previous studies. Many factors such as product formulation of enzyme, inappropriate method of providing the enzyme product to the animal, under- or over-supplementation of enzyme, diet composition, energy status of the target animals, and others are responsible for this high variability in response (Beauchemin et al. 2004a). On the other hand, the commercial enzyme products used for the ruminants are limited, and most of commercially available enzyme products been evaluated as ruminant feed additives are produced for non-feed application (Bhat and Hazlewood 2001), which also contributed to the high variability in response to enzyme supplementation in ruminants.

With the development of herbivorous animal husbandry, developing more effective enzyme products for ruminants is essential. In this chapter, the effectiveness of exogenous fibrolytic enzyme addition in ruminant's diet was evaluated using in vitro and in vivo studies that focused on determining the specific effect of source of enzyme, the enzyme activity, and methods of supplementation.

2 In Vitro Screening of Exogenous Fibrolytic Enzymes

2.1 *Materials and Methods*

2.1.1 Crop Straws, Grasses, and Enzymes

Three crop straws, that is, maize stover (Kexiangtian 1, MS), rice straw (Xiang 125s, RS), and wheat straw (Taishan 9818, WS), were obtained after the grain were harvested, and two types of grasses, that is, Guimu-1 (GM) at late tillering and alfalfa before flowering, were selected as substrates in this in vitro fermentation study. The selected crop straw and grasses were used as a substrate for screening candidate cellulase, and the maize stove, rice straw, and Guimu-1 were involved in the evaluating exogenous xylanase.

Table 1 Enzymatic activities (U)[a] and description of enzyme products (from He et al. 2015)

Enzyme[b]	Protein content (P)[c]	Xylanase (X)	Endoglucanase	X/P	Physical form
EPX1	154	8289	0	53.8	Powder
EPX2	284	20,768	0	70.1	Powder
EPX3	265	17,089	0	64.5	Powder
EPX4	128	651	0	5.10	Powder
EPX5	1150	66,945	143	58.2	Liquid

[a]One enzymatic unit was expressed as μmol xylose or glucose/min/g or mL enzyme from the corresponding substrate, using a standard curve made with xylose or glucose (from 0 to 0.8 mg/mL) obtained under identical conditions

[b]*EPX* enzyme product with high xylanase activity, *EPX1* xylanase gene of which was derived from a strain of *Trichoderma harzianum*, *EPX2* xylanase gene of which was derived from a strain of *Trichoderma reesei*, *EPX3* xylanase gene of which was derived from a strain of *Orpinomyces*, *EPX4* xylanase gene which was derived from a strain of *Aspergillus oryzae*, *EPX5* was produced from *Trichoderma longibrachiatum*

[c]Protein content of enzymes were expressed in terms of mg/g (EXP1-EXP4) or mL (EXP5)

Exogenous cellulases (ECs) were procured from *Neocallimastix patriciarum* (NP; Hunan Youtell Biochemical, Ltd. Yueyang, China), *Trichoderma Longibrachiatum* (TL; Wuhan Sunhy Biological Co., LTD, Wuhan, China), and *Trichoderma Reesei* (TR; Guangdong VTR Biological Technology Co., Ltd., Zhuhai, China), respectively.

Five xylanase products containing single enzyme activity were used. The enzyme products EPX1, EPX2, and EPX3 were extracted from *Trichoderma harzianum*, rumen bacteria, and rumen fungi, respectively, whereas the EPX4 and EPX5 were extracted from *Aspergillus oryzae*. The enzyme products EPX1, EPX2, EPX3, and EPX4 were in powder form, and EPX5 was in liquid form. The enzymatic characteristics of the EPX products are presented in Table 1.

2.1.2 In Vitro Gas Production and Sampling

In vitro fermentation technique for candidates cellulase and xylanase screening was conducted according to the procedures modified by Tang et al. (2006) and Goering and Van Soest (1970), respectively. The samples of gas, residues, and fermentation liquor were collected to measure the methane production, DM, OM, or fiber degradation and fermentation characteristics, as the forage was supplemented with different dosages of exogenous fibrolytic enzymes.

2.2 *Results and Discussion*

2.2.1 Gas Production Parameters

Effects of Exogenous Cellulase Addition For grasses, in vitro gas production parameters generally were not affected by EC source, except for *Vf*, which increased ($P < 0.001$) by 8% and 19% for grasses supplemented with NP than those supplemented with TL and TR, respectively, and the grasses supplemented with NP had the maximum *Vf* value (Table 2). Only *Vf* and *k* were influenced by the dose rate of supplemented EC. The effects of TR (linear, $P < 0.001$) and NP (quadratic, $P < 0.05$) inclusion on *Vf*, and TL (linear, $P = 0.01$) and NP (linear, $P = 0.006$) inclusion on *k* were dose dependent. The optimum dose rate (DR) for NP which was the minimum dose required to elicit the greatest significant increase in gas production and gas production rate compared with the control, being at 12 IU/g of DM.

For crop residue, the EC source affected *Vf* ($P < 0.001$) and *d* ($P = 0.02$) values, and the lowest *Vf* was observed in TR (Table 3). Compared with the TR, the TL and NP relative improvements in *Vf* were 17% and 9% . The effect of all EC addition on in vitro gas production parameters, except for *RFD0*, depended upon the DR used, as proved by significant EC × DR interactions. The linear effects on *Vf*, *k*, *b*, and *d* values to DR were found for TL, and linear effects on *Vf*, *d* and *T0.5* values to DR were also found for NP. A linear response on *Vf*, a quadratic response on *k* and *d* values to DR were observed for TR, respectively. The optimum DR varied among EC. Low DR of NP (12 IU/g of DM) and the highest DR of TL (62 IU/g of DM) increased *Vf* by 5% and 7.0% compared with control.

The maximum gas production of grass and crop straw differed among the three EC, but it was similar to that of Eun and Beauchemin (2007) and Chen et al. (2013). Eun and Beauchemin (2007) reported that feed enzymes originated from *Trichoderma longibrachiatum* had higher gas production than that from *Penicillum funiculosum*. In the present study, crop straw or grass supplemented with EC of TL and NP had higher gas production than that of TR. Also, Chen et al. (2013) found that gas production at 48 h incubation time ranged from 166 to 177 mL/g for corn silage added 5 cellulases. Composition of exoglucanase, endoglucanase, and xylanase, and their activity in the enzyme system, may be responsible for this difference. It indicated that the origin of EC should be considered when EC was used to improve in vitro fermentation of forage. When the gas production was fitted to the Wang et al. (2011) model, no pronounced improvement in the maximum gas production compared with control and this was consistent with the finding of previous studies (Colombatto et al. 2004; Eun and Beauchemin 2007; Chen et al. 2013). Index of *FRD0* and *T0.5* reflect the rate of fermentation at early incubation stages of "<12 h" and the incubation time of half of the maximum gas production reached, respectively. Supplementation of EC, although, did not increase the value of *T0.5* and *FRD0* obviously, a numerical decrease in *T0.5* and an increase in *FRD0* indicated that EC supplementation might be beneficial to improve the effect of bacteria on the degradation of forage. On the other hand, *T0.5* is more than 14 h for grass and 20 h for crop straw, indicating that the degradation period of bacteria on forage is mainly in the middle or late stages of incubation. These results also support the general agreement that enzymes increased the rate, but not the extent, of feed degradation in the rumen (Beauchemin et al. 2001).

Effects of Exogenous Xylanase Addition There were no interactions between EPX and dose for GP kinetics and other measurements for any of the forages; therefore, only the main effects of EPX and dose are discussed (Tables 1, 2, 3, and 4). For MS,

Table 2 In vitro fermentation characteristics of grass supplemented with exogenous cellulose (from Tang et al. 2013)

		Dose rate[3]						Significance of effect[5]		
Item[1]	Enzyme[2]	Mean	0	12	37	62	SEM[4]	EC	DR	EC × DR
Vf, mL	TL	274^{f}	281	266	272	284	8.4	$P < 0.001$	NS	$P < 0.001$
	TR	247^{g}	281^{a}	260^{ab}	245^{ab}	236^{b}			L($P < 0.001$)	
	NP	295^{e}	281	326	272	287			NS	
	SEM[6]	5.3								
k, /h	TL	0.065	0.059^{b}	0.058^{ab}	0.068^{a}	0.068^{a}	0.0051	NS	L($P = 0.01$)	NS
	TR	0.062	0.059	0.050	0.067	0.068			NS	
	NP	0.069	0.059^{ab}	0.061^{ab}	0.064^{ab}	0.082^{a}			L($P = 0.006$)	
	SEM[6]	0.0027								
b	TL	−1.78	−2.03	−2.31	−1.18	−1.86	1.333	NS	NS	NS
	TR	−1.33	−2.03	−3.73	−0.01	−0.25			NS	
	NP	−0.98	−2.03	−0.45	−3.11	0.63			NS	
	SEM[6]	0.659								
d	TL	−0.126	−0.183	−0.182	−0.087	−0.109	0.0487	NS	NS	NS
	TR	−0.106	−0.183	−0.056	−0.137	−0.126			NS	
	NP	−0.123	−0.183	0.001	−0.150	−0.221			NS	
	SEM[6]	0.0191								
T0.5, h	TL	15.3	16.3	15.7	15.3	14.7	1.13	NS	NS	NS
	TR	14.9	16.3	15.3	14.6	14.8			NS	
	NP	15.5	16.3	16.8	13.5	16.1			NS	
	SEM[6]	0.65								

FRD0, /h	TL	0.036	0.029	0.034	0.036	0.037	0.0052	NS	NS	NS
	TR	0.037	0.029	0.040	0.034	0.036			NS	
	NP	0.034	0.029	0.034	0.040	0.027			NS	
	SEM[6]	0.0029								

[e–g] Means within a column for EC that do not have a common superscript differ at $P < 0.05$

[a–b] Means within a row for dose rates of 0 to 62 IU/g of DM that do not have a common superscript differ at $P < 0.05$

[1] *Vf* = the maximum gas production; *k* = gas production fraction; *b* and *d* = the shape of the gas production curve; *T0.5* = the time when half of the maximum gas production reached; *FRD0* = the initial fractional rate of degradation

[2] TL, TR, and NP were cellulase originating from *Trichoderma longibrachiatum* (for TL), *Trichoderma reesei* (for TR) or *Neocallimastix patriciarum* (for NP)

[3] Dose rate as IU/g of DM forage substrate; Mean = mean for individual EC across dose rates except dose rate of 0; 0 = control without added EC

[4] SEM for EC × DR

[5] DR = dose rate; L = linear effect of DR; EC × DR = interaction between EC and DR; NS means $P > 0.10$

[6] SEM for pooled mean of EC excluding the dose rate of 0

Table 3 In vitro fermentation characteristics of straws supplemented with exogenous cellulase (from Tang et al. 2013)

Item[1]	Enzyme[2]	Dose rate[3]					SEM[4]	Significance of effect[5]		
		Mean	0	12	37	62		EC	DR	EC × DR
Vf, mL	TL	275^{e}	274^{ab}	243^{b}	290^{ab}	293^{a}	10.8	$P < 0.001$	L($P = 0.03$)	$P < 0.001$
	TR	235^{f}	274^{a}	263^{ab}	216^{b}	227^{ab}			L($P < 0.001$)	
	NP	259^{e}	274^{ab}	289^{a}	238^{b}	251^{ab}			L($P = 0.01$)	
	SEM[6]	6.1								
k, /h	TL	0.059	0.062^{a}	0.061^{ab}	0.054^{ab}	0.052^{b}	0.0036	NS	L($P = 0.04$)	$P < 0.001$
	TR	0.061	0.062^{b}	0.043^{c}	0.083^{a}	0.057^{b}			Q($P < 0.001$)	
	NP	0.061	0.062	0.057	0.061	0.065			NS	
	SEM[6]	0.0023								
b	TL	0.402	0.798^{a}	0.734^{a}	0.383^{ab}	0.091^{b}	0.2509	NS	L($P = 0.01$)	$P < 0.001$
	TR	0.392	0.798	−0.807	1.398	0.586			NS	
	NP	0.580	0.798	0.686	0.537	0.518			NS	
	SEM[6]	0.1631								
d	TL	-0.214^{e}	-0.147^{a}	-0.418^{b}	-0.114^{a}	-0.111^{a}	0.0528	$P = 0.02$	L($P < 0.001$)	$P < 0.001$
	TR	-0.358^{f}	-0.147^{a}	-0.153^{a}	-0.674^{b}	-0.248^{a}			Q($P < 0.001$)	
	NP	-0.259^{ef}	-0.147^{a}	-0.142^{a}	-0.242^{ab}	-0.393^{b}			L($P = 0.01$)	
	SEM[6]	0.0340								
T0.5, h	TL	21.4	22.0	20.6	21.8	21.7	0.73	NS	NS	$P = 0.01$
	TR	20.5	22.0	20.2	20.3	21.2			Q($P = 0.04$)	
	NP	20.7	22.0^{a}	23.3^{a}	19.7^{b}	19.2^{b}			L($P < 0.001$)	
	SEM[6]	0.45								

FRD0, /h	TL	0.021	0.019	0.019	0.022	0.022	0.0015	NS	NS	$P = 0.01$
	TR	0.020	0.019	0.024	0.016	0.020			NS	
	NP	0.021	0.019	0.019	0.022	0.022			NS	
	SEM[6]	0.0009								

[e-f]Means within a column for EC that do not have a common superscript differ at $P < 0.05$

[a-c]Means within a row for dose rates of 0 to 62 IU/g of DM that do not have a common superscript differ at $P < 0.05$

[1]*Vf* = the maximum gas production; *k* = gas production fraction; *b* and *d* = the shape of the gas production curve; *T0.5* = the time when half of the maximum gas production is reached; *FRD0* = the initial fractional rate of degradation

[2]TL, TR, and NP were cellulose originating from *Trichoderma longibrachiatum* (for TL), *Trichoderma Reesei* (for TR), or *Neocallimastix patriciarum* (for NP)

[3]Dose rate as IU/g of DM forage substrate; Mean = mean for individual EC across dose rates except dose rate of 0; 0 = control without added EC

[4]SEM for EC × DR

[5]DR = dose rate; L = linear effect of DR; EC × DR = interaction between EC and DR; NS means $P > 0.10$

[6]SEM for pooled mean of EC excluding the dose rate of 0

Table 4 Gas production kinetic characters for the three forages used in the current study[a] (from He et al. 2015)

Item		MS			RS			GM		
		Vf	*T0.5*	*FRD*0	*Vf*	*T0.5*	*FRD*0	*Vf*	*T0.5*	*FRD*0
EPX group[b]	EPX1	231	24.8	0.017	202	21.5	0.012	240	18.3	0.031
	EPX2	234	20.7	0.018	198	21.2	0.014	231	16.2	0.039
	EPX3	221	23.5	0.017	201	21.2	0.014	237	17.2	0.031
	EPX4	228	22.7	0.017	207	21.1	0.015	248	18.5	0.030
	EPX5	221	19.6	0.022	209	21.2	0.016	240	17.4	0.032
$S.E.M_{EPX}$		11.6	2.36	0.0037	6.7	0.74	0.0009	10.3	4.00	0.0174
Dose[c]	0	226	19.7	0.017	189	19.9	0.014	229	14.5	0.051
	20	219	22.5	0.019	206	21.8	0.013	242	17.7	0.029
	50	238	22.5	0.019	208	21.8	0.013	239	18.0	0.027
	80	237	24.4	0.019	211	21.4	0.015	246	19.8	0.023
$S.E.M_{Dose}$		11.7	2.19	0.0037	6.4	0.71	0.0008	10.2	3.96	0.0171
P-value[d]										
EPX		NS	NS	NS	NS	NS	<0.01	NS	NS	NS
Dose		NS	NS	NS	<0.01	<0.01	<0.01	0.02	0.01	0.02
$Dose_{Linear}$		NS	NS	NS	<0.01	0.03	0.10	<0.01	<0.01	0.01
$Dose_{Quadratic}$		NS	NS	NS	0.07	0.01	<0.01	NS	NS	NS
EPX × Dose		NS	NS	NS	NS	NS	NS	NS	NS	NS

[a]*Vf* is the asymptotic gas production (mL/g DM); *T0.5* is the time (h) when the gas production is half of the theoretical maximum gas production; and *FRD*0 (%/h) is the initial fractional rate of degradation; MS = maize stover; RS = rice straw; GM = Guimu NO.1 grass; EPX = enzyme product with high xylanase activity
[b]EPX1, xylanase gene of which was derived from a strain of *Trichoderma harzianum*; EPX2, xylanase gene of which was derived from a strain of *Trichoderma reesei*; EPX3, xylanase gene of which was derived from a strain of *Orpinomyces*; EPX4, xylanase gene of which was derived from a strain of *Aspergillus oryzae*; EPX5 was produced from *Trichoderma longibrachiatum*
[c]Dose as units of xylanase/g of DM forage substrate; one enzymatic unit was expressed as μmol xylose/min/g or mL enzyme from oat spelt xylan, using a standard curve made with xylose (from 0 to 0.8 mg/mL) obtained under identical conditions
[d]NS means $P > 0.10$; $Dose_{Linear}$ = linear effect of dose; $Dose_{Quadratic}$ = quadratic effect of dose

there was no effect of EPX or dose on any of the GP kinetic variables. For RS, there was a difference in the effect of EPX ($P < 0.01$) on *FRD*0. The EX5 treatment resulted in a greater *FRD*0 for RS than any other treatments. Increasing EPX dose increased ($P < 0.05$) *vf* and *T*0.5 for RS and GM. For RS, these GP kinetic variables were or tended to be linearly (*vf* and *T*0.5, $P < 0.05$; *FRD*0, $P = 0.096$) and quadratically (*vf*, $P = 0.065$; *T*0.5 and *FRD*0, $P < 0.05$) increased with increasing EPX dose, whereas for GM, they were only increased linearly ($P < 0.05$). Additionally, the *FRD*0 was linearly ($P < 0.05$) decreased for GM with increasing EPX dose.

Enzyme treatment exerted positive effects on GP variables. For example, the increase in *FRD*0 (in vitro rate of fermentation measured at early stages) with EPX addition indicated that EPX accelerated the rate of fermentation. However, this improvement response only occurred in RS but not in MS and GM, suggesting enzyme-forage specificity. Increased *FRD*0 would be beneficial in vivo because the rate of fermentation measured at the early stages of in vitro incubations was, to an extent, an indicator of the in vivo feed efficiency (Wang et al. 2013).

2.2.2 Dry Matter Degradability, Neutral Detergent Fiber Degradability, or Methane Production

Effects of Exogenous Cellulose Supplementation For grasses, the effect of each of 3 EC on dry matter degradability (DMD) depended on the source of EC and DR used, while for neutral detergent fiber degradability (NDFD), EC supplementing effects were only related to DR used (Table 5). The source of EC affected ($P < 0.001$) DMD, and the lowest DMD was noted in TL, which decreased 13.8% and 13.4% of NDFD compared with TR and NP, respectively. A quadratic response on DMD and NDFD to DR was observed for all EC except for NDFD of NP, which linearly ($P = 0.01$) increased the NDFD of grasses. Degradation of DM and neutral detergent fiber (NDF) of grasses were increased when EC was supplemented. Optimum DR for improving DMD and NDFD was 12 IU/g of DM for TL, TR and NP compared with control. For the TL, TR, and NP, relative improvements in DMD were 26.7%, 28.1% and 24.6%, and in NDFD were 17%, 24.4% and 24%, respectively.

For crop straws, the EC source only affected ($P < 0.001$) DMD (Table 5). The difference in NDFD among the three EC was not observed. Crop residues supplemented with EC from TR and NP increased DMD by 16.9% and 14.5% compared with those supplemented with EC from TL. All 3 EC influences on DMD and NDFD relied on the DR used, as evidenced by significant EC × DR interaction (Table 5). A quadratic response on DMD and NDFD to DR was observed for all EC. The optimum DR in increasing DMD was 12 IU/g of DM for 3 EC, and in improving NDFD was 12 IU/g of DM for TL and NP, and 37 IU/g of DM for TR. The lowest DR of TL, TR and NP (12 IU/g of DM) increased DMD by 21.2%, 19.2% and 15% compared with control. Relative improvements in the NDFD were 23.8% for TL, 25.7% for NP, and 31.7% for TR.

The DMD and NDFD data analysis revealed that enzyme treatments increased the degradable fraction of the forages, which is in agreement with the data obtained from grass (Zhu et al. 1999), alfalfa (Chen et al. 1994; Eun and Beauchemin 2007), corn silage (Eun and Beauchemin 2007; Sun et al. 2009; Chen et al. 2013), and with our previous report (Tang et al. 2008). Other studies using maize silage or corn silage have reported small increases (Chen et al. 1994; Eun and Beauchemin 2007). Our results suggest that more macromolecular were hydrolyzed to simpler and degradable ones as enzymes were supplemented in forages. Origin of EC differed in improving DMD and NDF was also observed in previous studies (Colombatto et al. 2004; Eun and Beauchemin 2007). The potential cause may be related to the activity of endoglucanase, which hydrolyzes cellulose chains at random, and exoglucanase, which hydrolyze cellulose chain from the nonreducing end (Bhat and Hazlewood 2001), in the enzyme system. Eun and Beauchemin (2007) also found endoglucanase linked to NDF degradability of alfalfa and corn silage. A significant increase in

Table 5 In vitro disappearance of DM and NDF of forage supplemented with exogenous cellulose (from Tang et al. 2013)

		Dose rate[3]						Significance of effect[5]		
Item[1]	Enzyme[2]	Mean	0	12	37	62	SEM[4]	EC	DR	EC × DR
Grass										
DMD, %	TL	47.2^{f}	35.1^{c}	61.8^{a}	39.3^{bc}	40.6^{b}	1.82	$P < 0.001$	Q($P = 0.01$)	$P < 0.001$
	TR	61.0^{e}	35.1^{b}	63.2^{a}	60.6^{a}	59.1^{a}			Q($P < 0.001$)	
	NP	60.6^{e}	35.1^{b}	59.7^{a}	59.1^{a}	63.0^{a}			Q($P < 0.001$)	
	SEM[6]	1.20								
NDFD, %	TL	76.8	57.6^{b}	74.6^{a}	76.1^{a}	79.7^{a}	4.18	NS	Q($P = 0.03$)	NS
	TR	83.9	57.6^{b}	82.0^{a}	87.6^{a}	82.3^{a}			Q($P < 0.001$)	
	NP	82.3	57.6^{c}	81.6^{ab}	75.5^{b}	89.7^{a}			L($P = 0.01$)	
	SEM[6]	2.46								
Straw										
DMD, %	TL	35.4^{f}	32.5^{b}	53.7^{a}	26.0^{c}	26.5^{c}	1.87	$P < 0.001$	Q($P = 0.02$)	$P < 0.001$
	TR	52.3^{e}	32.5^{b}	51.7^{a}	53.5^{a}	51.8^{a}			Q($P < 0.001$)	
	NP	49.9^{e}	32.5^{b}	47.5^{a}	51.2^{a}	50.9^{a}			Q($P < 0.001$)	
	SEM[6]	1.20								
NDFD, %	TL	60.3	39.4^{b}	63.2^{a}	61.9^{a}	55.7^{a}	3.30	NS	Q($P < 0.001$)	$P = 0.02$
	TR	59.4	39.4^{c}	55.2^{b}	65.1^{a}	58.0^{ab}			Q($P < 0.001$)	
	NP	66.2	39.4^{c}	71.1^{a}	58.6^{b}	69.0^{a}			Q($P = 0.02$)	
	SEM[6]	1.91								

[e–f]Means within a column for EC (exogenous cellulase) that do not have a common superscript differ at $P < 0.05$

[a–c]Means within a row for dose rates of 0 to 62 IU/g of DM that do not have a common superscript differ at $P < 0.05$

[1]DMD = DM degradability; NDFD = NDF degradability

[2]TL, TR, and NP were cellulose originating from *Trichoderma longibrachiatum* (for TL), *Trichoderma Reesei* (for TR), or *Neocallimastix patriciarum* (for NP)

[3]Dose rate as IU/g of DM forage substrate; Mean = mean for individual EC across dose rates except dose rate of 0; 0 = control without added EC

[4]SEM for EC × DR

[5]DR = dose rate; L = linear effect of DR; EC × DR = interaction between EC and DR; NS means $P > 0.10$

[6]SEM for pooled mean of EC excluding the dose rate of 0

DMD or NDFD indicated catalytic effects of EC on the substrate (Morgavi et al. 2001). In any case, the outcome indicated that the fermentation efficiency of forage might be improved as enzymes are used as additives in ruminant feedstuffs.

Effects of Exogenous Xylanase Supplementation For MS, there were no EPX group effects for in vitro dry matter degradability (IVDMD), in vitro neutral detergent fiber degradability (IVNDFD), CH_4 or CH_4/IVDMD at either incubation time

Table 6 Methane production and degradability coefficient of maize stover[a] (from He et al. 2015)

Item		24 h of incubation				48 h of incubation			
		IVDMD	IVNDFD	CH_4	CH_4/IVDMD	IVDMD	IVNDFD	CH_4	CH_4/IVDMD
EPX group[b]	EPX1	0.434	0.273	11.6	26.7	0.569	0.536	31.7	55.9
	EPX2	0.434	0.237	9.9	23.0	0.570	0.522	28.4	50.1
	EPX3	0.422	0.221	10.8	25.6	0.570	0.531	27.0	47.8
	EPX4	0.434	0.239	12.5	28.8	0.555	0.510	27.0	48.9
	EPX5	0.431	0.275	13.7	31.9	0.565	0.521	28.7	51.0
$S.E.M_{EPX}$		0.0320	0.0089	1.15	2.65	0.0068	0.0088	1.44	2.90
Dose[c]	0	0.413	0.178	12.6	30.7	0.576	0.521	31.0	53.9
	20	0.434	0.252	11.1	25.6	0.565	0.529	27.3	48.6
	50	0.435	0.272	11.7	27.1	0.565	0.522	28.6	51.1
	80	0.443	0.300	11.3	25.4	0.557	0.525	27.3	49.3
$S.E.M_{Dose}$		0.0286	0.0080	1.03	2.37	0.0061	0.0078	1.29	2.60
P-value[d]									
EPX		NS	NS	NS	NS	NS	NS	NS	NS
Dose		0.07	<0.01	NS	NS	NS	NS	NS	NS
$Dose_{Linear}$		0.02	<0.01	NS	NS	0.06	NS	NS	NS
$Dose_{Quadratic}$		NS	NS	NS	NS	NS	NS	NS	NS
EPX × Dose		NS	NS	NS	NS	NS	0.08	NS	NS

[a]GP = gas production (mL/g DM); IVDMD = in vitro degradability of DM; IVNDFD = in vitro degradability of NDF; CH_4 = methane (mL/g DM incubated); CH_4/IVDMD = methane/in vitro degradability of DM (mL/g DM digested); EPX = enzyme product with high xylanase activity
[b]EPX1, xylanase gene of which was derived from a strain of *Trichoderma harzianum*; EPX2, xylanase gene of which was derived from a strain of *Trichoderma reesei*; EPX3, xylanase gene of which was derived from a strain of *Orpinomyces*; EPX4, xylanase gene of which was derived from a strain of *Aspergillus oryzae*; EPX5 was produced from *Trichoderma longibrachiatum*
[c]Dose as units of xylanase/g of DM forage substrate; one enzymatic unit was expressed as μmol xylose/min/g or mL enzyme from oat spelt xylan, using a standard curve made with xylose (from 0 to 0.8 mg/mL) obtained under identical conditions
[d]NS means $P > 0.10$; $Dose_{Linear}$ = linear effect of dose; $Dose_{Quadratic}$ = quadratic effect of dose

(Table 6). There was a linear response to increasing EPX dose for IVDMD after 24 ($P < 0.05$) and 48 h ($P = 0.063$) and IVNDFD ($P < 0.01$) only at 24 h. However, CH_4 production was not affected by EPX dose.

For RS, dose increased ($P < 0.01$) CH_4 production at 24 h in a linear manner, with no EPX group effects (Table 7). However, these same variables were not affected by treatments at 48 h. There were effects ($P < 0.01$) of EPX on IVDMD and IVNDFD at 24 h, but not at 48 h. The degradability coefficients of DM and NDF recorded for EPX4 group (0.373 and 0.250, respectively) were the highest, while EPX1 group had the lowest (0.344 and 0.207, respectively). Increasing EPX dose increased (linear effect, $P < 0.01$; quadratic effect, $P < 0.01$) IVDMD and IVNDFD at 24 h, but only IVNDFD was affected by dose at 48 h (linear effect, $P < 0.01$; quadratic effect, $P < 0.01$).

Table 7 Methane production and degradability coefficient of rice straw[a] (from He et al. 2015)

Item		24 h of incubation				48 h of incubation			
		IVDMD	IVNDFD	CH_4	CH_4/IVDMD	IVDMD	IVNDFD	CH_4	CH_4/IVDMD
EPX group[b]	EPX1	0.344	0.207	10.7	31.0	0.512	0.445	25.1	49.0
	EPX2	0.358	0.225	11.5	32.2	0.510	0.445	24.4	48.0
	EPX3	0.364	0.219	11.5	31.6	0.513	0.446	25.0	48.9
	EPX4	0.373	0.250	11.5	30.8	0.507	0.456	25.7	50.9
	EPX5	0.367	0.242	11.0	30.0	0.511	0.458	26.7	52.1
$S.E.M_{EPX}$		0.0050	0.0068	0.37	0.97	0.0053	0.0065	0.62	1.27
Dose[c]	0	0.333	0.182	10.1	30.2	0.505	0.427	25.9	51.4
	20	0.371	0.246	11.4	30.7	0.517	0.457	25.4	49.2
	50	0.368	0.238	11.7	32.0	0.510	0.462	24.7	48.6
	80	0.372	0.248	11.8	31.7	0.510	0.454	25.4	50.0
$S.E.M_{Dose}$		0.0044	0.0061	0.33	0.87	0.0047	0.0058	0.56	1.13
P-value[d]									
EPX		<0.01	<0.01	NS	NS	NS	NS	NS	NS
Dose		<0.01	<0.01	<0.01	NS	NS	<0.01	NS	NS
$Dose_{Linear}$		<0.01	<0.01	<0.01	NS	NS	<0.01	NS	NS
$Dose_{Quadratic}$		<0.01	<0.01	NS	NS	NS	<0.01	NS	NS
EPX × Dose		NS	NS	NS	NS	NS	NS	NS	NS

[a]GP = gas production (mL/g DM); IVDMD = in vitro degradability of DM; IVNDFD = in vitro degradability of NDF; CH_4 = methane (mL/g DM incubated); CH_4/IVDMD = methane/in vitro degradability of DM (mL/g DM digested); EPX = enzyme product with high xylanase activity
[b]EPX1, xylanase gene of which was derived from a strain of *Trichoderma harzianum*; EPX2, xylanase gene of which was derived from a strain of *Trichoderma reesei*; EPX3, xylanase gene of which was derived from a strain of *Orpinomyces*; EPX4, xylanase gene of which was derived from a strain of *Aspergillus oryzae*; EPX5 was produced from *Trichoderma longibrachiatum*
[c]Dose as units of xylanase/g of DM forage substrate; one enzymatic unit was expressed as μmol xylose/min/g or mL enzyme from oat spelt xylan, using a standard curve made with xylose (from 0 to 0.8 mg/mL) obtained under identical conditions
[d]NS means $P > 0.10$; $Dose_{Linear}$ = linear effect of dose; $Dose_{Quadratic}$ = quadratic effect of dose

For GM, there were no effects of EPX on degradability or CH_4 variables, except for IVNDFD coefficient after 48 h of incubation, which was greater ($P < 0.05$) with EPX5 (0.628) than with other EPX (Table 8). Increasing EPX dose increased IVNDFD (linear effect, $P < 0.01$; quadratic effect, $P < 0.01$), CH_4 (linear effect, $P < 0.01$) and CH_4/IVDMD (linear effect, $P < 0.05$) at 24 h, but there was no effect of dose on these variables at 48 h.

Supplementation of EPX products substantially improved the degradation of MS, RS, and GM. For instance, at the dose of 20 units/g DM, the EPX products increased 24-h IVNDFD of MS by 41.6%, RS by 35.2%, and GM by 47.1% when compared to the control. The improvement of IVNDFD can allow greater voluntary feed intake (Dado and Allen 1995) by ruminants due to decreased rumen fill, and it can also increase the energy density of diets and stimulate microbial nitrogen production (Oba and Allen 2000). It has been proposed that a 0.01 increase in forage

Table 8 Methane production and degradability coefficient of Guimu No. 1 grass[a] (from He et al. 2015)

Item		24 h of incubation				48 h of incubation			
		IVDMD	IVNDFD	CH_4	CH_4/IVDMD	IVDMD	IVNDFD	CH_4	CH_4/IVDMD
EPX group[b]	EPX1	0.464	0.370	14.4	31.1	0.615	0.598	30.9	50.3
	EPX2	0.477	0.371	14.7	30.9	0.626	0.606	26.6	42.5
	EPX3	0.463	0.354	14.6	31.6	0.631	0.603	29.9	47.4
	EPX4	0.465	0.352	14.9	32.2	0.631	0.599	30.4	48.3
	EPX5	0.467	0.352	15.3	32.6	0.636	0.628	29.4	46.2
$S.E.M_{EPX}$		0.0076	0.0131	0.63	1.31	0.0050	0.0095	1.63	2.53
Dose[c]	0	0.469	0.263	14.2	30.2	0.630	0.603	29.5	46.9
	20	0.462	0.387	13.9	30.2	0.629	0.612	29.1	46.4
	50	0.466	0.393	15.2	32.7	0.624	0.603	28.7	45.8
	80	0.471	0.398	15.8	33.6	0.628	0.609	30.5	48.6
$S.E.M_{Dose}$		0.0068	0.0117	0.57	1.18	0.0043	0.0089	1.52	2.32
P-value[d]									
EPX		NS	NS	NS	NS	NS	0.03	NS	NS
Dose		NS	<0.01	0.08	0.10	NS	NS	NS	NS
$Dose_{Linear}$		NS	<0.01	<0.01	0.02	NS	NS	NS	NS
$Dose_{Quadratic}$		NS	<0.01	NS	NS	NS	NS	NS	NS
EPX × Dose		NS	NS	NS	NS	NS	NS	NS	NS

[a]GP = gas production (mL/g DM); IVDMD = in vitro degradability of DM; IVNDFD = in vitro degradability of NDF; CH_4 = methane (mL/g DM incubated); CH_4/IVDMD = methane/in vitro degradability of DM (mL/g DM digested); EPX = enzyme product with high xylanase activity
[b]EPX1, xylanase gene of which was derived from a strain of *Trichoderma harzianum*; EPX2, xylanase gene of which was derived from a strain of *Trichoderma reesei*; EPX3, xylanase gene of which was derived from a strain of *Orpinomyces*; EPX4, xylanase gene of which was derived from a strain of *Aspergillus oryzae*; EPX5 was produced from *Trichoderma longibrachiatum*
[c]Dose as units of xylanase/g of DM forage substrate; one enzymatic unit was expressed as μmol xylose/min/g or mL enzyme from oat spelt xylan, using a standard curve made with xylose (from 0 to 0.8 mg/mL) obtained under identical conditions
[d]NS means $P > 0.10$; $Dose_{Linear}$ = linear effect of dose; $Dose_{Quadratic}$ = quadratic effect of dose

IVNDFD would elicit a 0.17 kg increase in DM intake and a 0.25 kg increase in the 4% fat-corrected milk yield of dairy cows (Oba and Allen 1999). Thus, using EPX as a feed additive may have the potential to improve the productivity of ruminants fed locally grown low-quality forages. Improvement of fiber degradation with EPX has resulted from enhanced attachment and colonization of plant cell wall material by rumen microorganisms (Nsereko et al. 2002), increased microbial growth (Eun et al. 2007), or synergism with endogenous microbial enzymes (Morgavi et al. 2001).

The dose response of EPX in increasing IVDMD and IVNDFD was forage dependent, indicating that the effectiveness of a particular enzyme dose differed among the three forages. However, the lack of interactions between EPX and dose for the three forages indicated that the dose response was not product dependent within the forage. The linear increases in IVDMD or IVNDFD with increasing EPX dose for MS indicated no inhibitory effects of EPX dose on xylan degradation. However, the quadratic responses to EPX dose for RS and GM indicated diminishing effects with increasing dose rate and suggested that a high rate of EPX may have inhibitory effects on xylan degradation by blocking the enzyme-binding sites (Nsereko et al. 2000) or decreasing microbial attachment to substrate (Morgavi et al. 2004). Such effects were also observed in vitro by Eun and Beauchemin (2008) and in vivo with other substrates and enzyme products (Nsereko et al. 2002). Based on the degradation results, it was concluded that, among the three forages evaluated, MS showed the most potential for improvement of fiber degradation using EPX.

Four of the xylanase products (except EPX5) were single activity products derived from different organism sources, but assessed at equal xylanase activity to explore the effectiveness of variable sources of EPX on GP and forage degradation. The greater IVDMD and IVNDFD of RS using EPX4 (i.e., derived from *Aspergillus oryzae*) indicated that EPX source is a factor that influences the efficacy of exogenous xylanases on forage degradation. Similar results for maize silage were also reported by Eun and Beauchemin (2008), in which increases in IVDMD and IVNDFD were observed after applying different sources of enzymes based on equal xylanase activity, with very low endoglucanase activity. However, the specific sources of the enzymes used were not mentioned. The reason for the differences in effects of EPX from different sources is not certain. It was expected that EPX4 (from *Aspergillus oryzae*) would have maximal synergistic effects in rumen fluid, because it can increase the number of ruminal bacteria (Newbold et al. 1992a, b). Furthermore, the fungus *Aspergillus oryzae* has been shown to produce an esterase capable of breaking ester bridges form between ferulic and *p*-coumaric acids form and arbinoxylan (Tenkanen et al. 1991), and the extent of cross-linking by *p*-coumaryl and feruloyl groups to arabinoxylans is one factor that limits the digestion of plant cell walls (McAllister et al. 2001). Moreover, it must be stated that xylanase activity was determined using oat spelt xylan as substrate and under a specific conditions (pH 6.0 and 39 °C) similar to those observed in the present incubation. However, the activity could have been different if the testing conditions changed, as activity depends on the conditions used (Colombatto and Beauchemin 2003). However, not all the responses to different EPX sources were positive. The minimal difference in MS and GM degradation among EPX suggested the responses to different EPX sources were also forage-dependent.

2.2.3 pH Value and CH_4 Production

The pH value and CH_4 production of grass were influenced ($P < 0.001$) by the source of EC (Table 9). Grasses supplemented with EC of TR and NP had lower pH value and CH_4 production than those supplemented with EC of TL. The decreases in pH were 3.8% and 3.2%, and in CH_4 were 44% and 24% for TR and NP compared with TL. Dosage of each three EC influenced pH value and CH_4 production of grasses. The EC of TR and NP decreased 4 times, while EC of TL increased 4 times of pH value and CH_4 production of grasses compared with control,

Table 9 Methane production and pH value of fermentation liquor of forage supplemented with exogenous cellulase (adapted from Tang et al. 2013)

Item[1]	Enzyme[2]	Dose rate[3]					SEM[4]	Significance of effect[5]		
		Mean	0	12	37	62		EC	DR	EC × DR
Grass										
pH	TL	6.80^{e}	6.93^{a}	6.50^{b}	6.94^{a}	6.96^{a}	0.023	$P < 0.001$	Q($P < 0.001$)	$P < 0.001$
	TR	6.54^{f}	6.93^{a}	6.49^{b}	6.58^{b}	6.55^{b}			Q($P < 0.001$)	
	NP	6.58^{f}	6.93^{a}	6.64^{b}	6.55^{b}	6.56^{b}			Q($P < 0.001$)	
	SEM[6]	0.013								
CH_4, mmol/g DMD	TL	5.20^{e}	6.19^{a}	3.23^{b}	6.28^{a}	6.08^{a}	0.428	$P < 0.001$	Q($P < 0.001$)	$P < 0.001$
	TR	2.91^{g}	6.19^{a}	3.03^{b}	2.90^{b}	2.81^{b}			Q($P < 0.001$)	
	NP	3.94^{f}	6.19^{a}	4.28^{b}	3.83^{b}	3.71^{b}			Q($P = 0.02$)	
	SEM[6]	0.219								
Straw										
pH	TL	6.77^{e}	6.87^{a}	6.44^{b}	6.94^{a}	6.93^{a}	0.019	$P < 0.001$	Q($P < 0.001$)	$P < 0.001$
	TR	6.53^{g}	6.87^{a}	6.48^{b}	6.54^{b}	6.57^{b}			Q($P < 0.001$)	
	NP	6.60^{f}	6.87^{a}	6.73^{b}	6.52^{c}	6.54^{c}			Q($P < 0.001$)	
	SEM[6]	0.011								
CH_4, mmol/g DMD	TL	7.20^{e}	7.75^{a}	3.31^{b}	9.15^{a}	9.13^{a}	0.518	$P < 0.001$	L($P < 0.001$)	$P < 0.001$
	TR	3.07^{g}	7.75^{a}	3.47^{b}	2.79^{b}	2.97^{b}			Q($P < 0.001$)	
	NP	3.90^{f}	7.75^{a}	4.45^{b}	3.71^{b}	3.55^{b}			Q($P < 0.001$)	
	SEM[6]	0.239								

[e–g]Means within a column for EC that do not have a common superscript differ at $P < 0.05$

[a–c]Means within a row for dose rates of 0 to 62 IU/g of DM that do not have a common superscript differ at $P < 0.05$

[1]CH_4 = methane production

[2]TL, TR, and NP were cellulose originating from *Trichoderma longibrachiatum* (for TL), *Trichoderma Reesei* (for TR) or *Neocallimastix patriciarum* (for NP)

[3]Dose rate as IU/g of DM forage substrate; Mean = mean for individual EC across dose rates except dose rate of 0; 0 = control without added EC

[4]SEM for EC × DR

[5]DR = dose rate; L = linear effect of DR; EC × DR = interaction between EC and DR

[6]SEM for pooled mean of EC excluding the dose rate of 0

respectively. The optimum DR of these three EC in decreasing pH and CH_4 was 12 IU/g of DM for TL, TR and NP. Relative decreases in pH and CH_4 were 6.2% and 48% for TL, 6.3% and 51% for TR, and 4.2% and 31% for NP, respectively.

For crop residue, the differences in EC source on pH value ($P < 0.001$) and CH_4 ($P < 0.001$) were observed among three EC (Table 9). The favorable EC in reducing pH and CH_4 was TR and NP, which decreased pH by 3.7% and 2.6%, and reduced CH_4 by 134% and 85% compared with TL, respectively. Dose rate response in pH and CH_4 varied among three EC, as evidenced by significant EC × DR interaction. Optimum DR for decreasing pH and CH_4 was 12 IU/g of DM for TL and TR, and 62 IU/g DM for NP was favorable. Relative decreases in CH_4 at this DR were 57%, 55%, and 54%, and in pH at those DR were 6.3%, 5.7% and 4.8% for TL, TR, and NP compared with control, respectively.

The pH value is a main factor that reflects internal environment of rumen. Treatments of TR and NP or EC reduced the pH value of fermentation liquor to 6.49 and 6.73 compared with TL, or control, suggesting that TR and NP, or EC treatments could maintain more suitable conditions for fermentation, and suitable for the growth of microorganisms. The final pH value of EC treatment ranged from 6.53 to 6.80, and can be considered optimal for fiber degradation in the rumen (Stewart et al. 1997). TR and NP or EC treatments in forage had higher DMD than TL or control, which may partially be explained by this reason. Satter and Slyter (1974) suggested that the lowest ammonia nitrogen (NH_3-N) concentration of rumen liquor should be higher than 5 mL/dL for bacteria to get the highest growth rate.

In the present study, EC treatment or EC of TR and NP treatments significantly decreased methane production compared with control or TL. Giraldo et al. (2007) found that methane production was not influenced by cellulase sprayed into the diet, which was composed of forage and concentration at 70:30. Whereas, Chung et al. (2012) reported that methane production, calculated as per kg of DM or kg of milk, would increase when enzymes were added to the dairy diet at 0.5 and 1.0 mL/kg. Dong et al. (1999) also found methane production increased by 43% when cellulase and xylanase were added to hay. Beauchemin et al. (2009) reported that enzyme supplementation though absolutely increased methane production, methane production per kg of milk would decrease. Methane production decrease may be related to microflora change of methanogenium led by enzyme addition (Zhou et al. 2011). Addition of EC decreased methane production, suggesting that utilization of cellulase, especially for TR and NP, in ruminant diets may be an efficient method of reducing the greenhouse effect caused by CH_4 emitted from ruminant production.

2.2.4 NH_3-N Concentration and Volatile Fatty Acid Profiles

Effects of Exogenous Cellulose Involving The source of EC influenced the concentration of NH_3-N in fermentation liquor ($P = 0.001$) of grasses (Table 10). Grasses supplemented with EC of TR and NP had lower NH_3-N concentrations than those supplemented with EC of TL. The decreases in NH_3-N were 28% and 39% for TR and NP compared with TL. Dosage of each EC influenced NH_3-N production of

Table 10 Volatile fatty acid profiles and NH_3-N concentrations in incubation solution of grasses supplemented with exogenous cellulase (adapted from Tang et al. 2013)

Item[1]	Enzyme[2]	Dose rate[3]					SEM[4]	Significance of effect[5]		
		Mean	0	12	37	62		EC	DR	EC × DR
NH_3-N, mg/L	TL	155^{e}	195^{a}	69^{b}	196^{a}	199^{a}	16.9	$P = 0.01$	L($P = 0.03$)	$P = 0.01$
	TR	111^{f}	195^{a}	93^{b}	102^{b}	138^{b}			Q($P < 0.001$)	
	NP	95^{f}	195^{a}	57^{c}	115^{b}	114^{b}			Q($P = 0.01$)	
	SEM[6]	10.4								
TVFA	TL	18.1^{f}	17.4	18.2	16.3	19.8	1.49	$P < 0.001$	NS	$P < 0.001$
	TR	19.1^{f}	17.4^{ab}	19.8^{ab}	20.5^{a}	17.1^{b}			Q($P = 0.02$)	
	NP	23.3^{e}	17.4^{c}	18.7^{bc}	28.1^{a}	23.2^{ab}			Q($P = 0.02$)	
	SEM[6]	0.95								
Acetate	TL	6.35^{f}	5.81	6.49	5.69	6.86	0.776	$P < 0.001$	NS	$P = 0.004$
	TR	7.30^{f}	5.81^{b}	7.36^{ab}	8.21^{a}	6.35^{b}			Q($P = 0.003$)	
	NP	9.57^{e}	5.81^{c}	7.90^{b}	11.95^{a}	8.86^{b}			Q($P = 0.003$)	
	SEM[6]	0.497								
Propionate	TL	5.74^{f}	5.56	6.17	4.93	6.13	0.453	$P = 0.007$	NS	$P = 0.01$
	TR	6.66^{ef}	5.56^{b}	6.34^{b}	7.51^{a}	6.14^{b}			Q($P = 0.002$)	
	NP	7.22^{e}	5.56^{b}	6.48^{ab}	7.71^{a}	7.46^{a}			L($P = 0.02$)	
	SEM[6]	0.296								
Butyrate	TL	3.46^{e}	3.38^{ab}	3.18^{b}	3.27^{b}	3.92^{a}	0.219	$P < 0.001$	L($P = 0.03$)	$P < 0.001$
	TR	2.79^{f}	3.38^{a}	3.62^{a}	2.49^{b}	2.25^{b}			L($P < 0.001$)	
	NP	3.80^{e}	3.38^{b}	2.45^{c}	4.92^{a}	4.02^{b}			L($P < 0.001$)	
	SEM[6]	0.138								
Isobutyrate	TL	0.788	0.825^{ab}	0.684^{b}	0.750^{b}	0.929^{a}	0.0451	NS	Q($P = 0.004$)	$P < 0.001$
	TR	0.775	0.825	0.665	0.819	0.840			NS	
	NP	0.759	0.825^{b}	0.460^{c}	0.990^{a}	0.827^{b}			L($P = 0.003$)	

(continued)

Table 10 (continued)

Item[1]	Enzyme[2]	Dose rate[3]					SEM[4]	Significance of effect[5]		
		Mean	0	12	37	62		EC	DR	EC × DR
	SEM[6]	0.0273								
Isovalerate	TL	0.933^{ef}	0.975^{ab}	0.839^{b}	0.880^{b}	1.079^{a}	0.0673	$P = 0.02$	Q($P = 0.009$)	$P < 0.001$
	TR	0.803^{f}	0.975^{a}	0.976^{a}	0.680^{b}	0.754^{b}			L($P = 0.002$)	
	NP	0.973^{e}	0.975^{b}	0.606^{c}	1.311^{a}	1.003^{b}			L($P = 0.01$)	
	SEM[6]	0.0409								
Valerate	TL	0.839^{ef}	0.836	0.820	0.771	0.925	0.0717	$P = 0.01$	NS	$P = 0.005$
	TR	0.791^{f}	0.836	0.872	0.744	0.758			NS	
	NP	0.987^{e}	0.836^{b}	0.804^{b}	1.170^{a}	0.987^{ab}			L($P = 0.04$)	
	SEM[6]	0.0452								
A:P	TL	1.08^{f}	1.04^{ab}	1.00^{b}	1.14^{a}	1.12^{a}	0.062	$P = 0.003$	L($P = 0.03$)	$P < 0.001$
	TR	1.07^{f}	1.04	1.10	1.10	1.01			NS	
	NP	1.25^{e}	1.04^{b}	1.19^{ab}	1.43^{a}	1.14^{b}			Q($P = 0.003$)	
	SEM[6]	0.036								

[e–f]Means within a column for EC that do not have a common superscript differ at $P < 0.05$

[a–c]Means within a row for dose rates of 0 to 62 IU/g of DM that do not have a common superscript differ at $P < 0.05$

[1]TVFA = Total volatile fatty acid; A:P = ration of acetic acid to propionic acid

[2]TL, TR, and NP were cellulose originating from *Trichoderma longibrachiatum* (for TL), *Trichoderma reesei* (for TR), or *Neocallimastix patriciarum* (for NP)

[3]Dose rate as IU/g of DM forage substrate; Mean = mean for individual EC across dose rates except dose rate of 0; 0 = control without added EC

[4]SEM for EC × DR

[5]EC = exogenous cellulase; DR = dose rate; L = linear effect of DR; Q = quadratic effect of DR; NS = no significant; EC × DR = interaction between EC and DR; NS means $P > 0.10$

[6]SEM for pooled mean of EC excluding the dose rate of 0

grasses. The EC of TR and NP decreased 4 times, while EC of TL increased 4 times of NH_3-N production of grasses compared with control.

For grasses, the EC source affected the concentration of VFA and the ratio of AA to PP in fermentation liquor, except for iso-butyrate (Table 10). The supplementation of grasses with EC of NP had the lowest NH3-N concentration and the highest proportion of TVFA, acetate, propionate, butyrate, iso-butyrate, valerate, iso-valerate, and the ratio of acetate to propionate. The EC of NP improved TVFA by 29% and 22%, acetate by 51% and 31%, and acetate to propionate ratio by 16% and 17% compared with TL and TR, respectively. The EC of NP also enhanced butyrate, isovalerate and valerate by 36%, 22% and 25% compared with TR and propionate by 26% compared with TL. The EC produced variable effects on short chain fatty acid across DR, indicated by the EC × DR interaction. The optimum DR for greater TVFA and individual volatile fatty acid were 37 IU/g of DM for NP compared with control, and relative improvements in these values at this DR ranged from 20% of isobutyrate to 106% of acetate. Optimum DR for increasing acetate and propionate was 37 IU/g of DM for TR compared with control. Relative improvements in these values at this DR were 41% and 35%. The EC of TL supplementation did not increase the concentration of TVFA, each volatile fatty acid, compared to the control.

For crop straws, the effects of EC source on NH_3-N ($P = 0.005$) were observed among three EC (Table 11). Compared with TL, the favorable EC in reducing NH_3-N was TR and NP, which decreased NH_3-N by 44% and 47%, respectively. Dose rate response in NH_3-N varied among EC, as evidenced by significant EC × DR interaction. Compared with control, the optimum DR for decreasing NH_3-N was 12 IU/g of DM for TL, TR, and NP. Relative decreases in NH_3-N at this DR were 57%, 55%, and 43% for TL, TR, and NP, respectively.

The source of EC had significant effects on TVFA ($P < 0.001$), acetate ($P < 0.001$), butyrate ($P < 0.001$), isovalerate ($P = 0.01$), and the value of A:P ($P < 0.001$) (Table 11). The highest TVFA, acetate, butyrate, and the ratio of acetate to propionate were observed in NP, and the highest isovalerate was noted in the EC of TR. The EC of NP increased ($P < 0.001$) TVFA by 21% and 23%, acetate by 47% and 37% compared with TL and TR, respectively, and propionate by 14% compared with TL. Compared with the EC of TL, supplementation of crop residues with EC of TR and NP elevated the production of butyrate and isovalerate by 28%, 34%, 20% and 19%, respectively. The EC produced variable effects on short chain fatty acid across DR, indicated by the EC × DR interaction, whereas the improvements in these parameters were very weak for crop residues supplemented with EC of TL and TR compared with control. The EC of NP linearly increased ($P < 0.001$) TVFA, butyrate, isovalerate, valerate, and the ratio of acetate to propionate, and quadratically increased acetate ($P = 0.02$) and propionate ($P = 0.04$) compared with control. Optimum DR for increasing the value of these parameters was 37 IU/g of DMD, except for A:P. The relative improvement in the values of these parameters at this DR were ranged from 21% of valerate to 75% of acetate compared with control. The EC supplementation could not decrease the ratio of acetate to propionate.

Table 11 Volatile fatty acid profiles and NH_3-N concentrations in incubation solution of straws supplemented with exogenous cellulase (adapted from Tang et al. 2013)

Item[1]	Enzyme[2]	Dose rate[3]					SEM[4]	Significance of effect[5]		
		Mean	0	12	37	62		EC	DR	EC × DR
NH_3-N mg/L	TL	143^{e}	227^{a}	74^{c}	202^{a}	153^{b}	16.6	$P = 0.005$	Q($P < 0.001$)	$P < 0.001$
	TR	99^{f}	227^{a}	109^{b}	89^{b}	98^{b}			Q($P < 0.001$)	
	NP	97^{f}	227^{a}	76^{b}	96^{b}	119^{b}			Q($P < 0.001$)	
	SEM[6]	6.4								
TVFA	TL	14.9^{f}	14.7	17.8	13.3	13.8	1.01	$P < 0.001$	NS	$P < 0.001$
	TR	15.1^{f}	14.7	16.2	14.3	14.8			NS	
	NP	18.3^{e}	14.7^{b}	14.3^{b}	21.3^{a}	19.4^{a}			L($P < 0.001$)	
	SEM[6]	0.61								
Acetate	TL	5.53^{f}	4.98	6.49	4.90	5.20	0.471	$P < 0.001$	NS	$P < 0.001$
	TR	5.16^{f}	4.98	5.66	4.88	4.95			NS	
	NP	7.60^{e}	4.98^{b}	5.84^{b}	8.71^{a}	8.25^{a}			Q($P = 0.02$)	
	SEM[6]	0.288								
Propionate	TL	5.21	4.91	5.95	4.61	5.05	0.333	NS	NS	$P = 0.02$
	TR	4.88	4.91^{ab}	5.65^{a}	4.37^{b}	4.62^{b}			L($P = 0.02$)	
	NP	5.57	4.91^{b}	5.25^{ab}	6.17^{a}	5.28^{ab}			Q($P = 0.04$)	
	SEM[6]	0.204								
Butyrate	TL	2.33^{f}	2.81^{a}	3.26^{a}	1.91^{b}	1.81^{b}	0.169	$P < 0.001$	L($P < 0.001$)	$P < 0.001$
	TR	2.98^{e}	2.81	3.03	2.93	2.97			NS	
	NP	3.13^{e}	2.81^{b}	2.02^{c}	3.90^{a}	3.46^{ab}			L($P < 0.001$)	
	SEM[6]	0.093								
Isobutyrate	TL	0.647	0.677	0.595	0.687	0.660	0.0337	NS	NS	$P < 0.001$
	TR	0.649	0.677^{a}	0.549^{b}	0.676^{a}	0.722^{a}			Q($P = 0.03$)	
	NP	0.646	0.677^{a}	0.363^{b}	0.780^{a}	0.795^{a}			L($P < 0.001$)	
	SEM[6]	0.0185								

Item[1]	Enzyme[2]	Dose rate[3]					SEM[4]	Significance of effect[5]		
		Mean	0	12	37	62		EC	DR	EC × DR
Isovalerate	TL	0.645[f]	0.714[ab]	0.822[a]	0.594[bc]	0.518[c]	0.0425	$P = 0.01$	NS	$P < 0.001$
	TR	0.771[e]	0.714	0.730	0.772	0.810			NS	
	NP	0.766[e]	0.714[b]	0.424[c]	0.976[a]	0.897[a]			L($P < 0.001$)	
	SEM[6]	0.0243								
Valerate	TL	0.573	0.636	0.637	0.559	0.523	0.0362	NS	NS	$P < 0.001$
	TR	0.642	0.636[ab]	0.562[b]	0.670[a]	0.693[a]			L($P = 0.008$)	
	NP	0.621	0.636[b]	0.399[c]	0.770[a]	0.694[ab]			L($P < 0.001$)	
	SEM[6]	0.0204								
A:P	TL	1.06[f]	0.995	1.06	1.09	1.02	0.046	$P < 0.001$	NS	$P < 0.001$
	TR	1.05[f]	0.995[ab]	0.984[b]	1.11[a]	1.06[ab]			L($P = 0.02$)	
	NP	1.29[e]	0.995[b]	1.05[b]	1.38[a]	1.45[a]			L($P < 0.001$)	
	SEM[6]	0.025								

[e–f]Means within a column for EC that do not have a common superscript differ at $P < 0.05$
[a–c]Means within a row for dose rates of 0 to 62 IU/g of DM that do not have a common superscript differ at $P < 0.05$
[1]SCFA = short chain fatty acid; A:P = ratio of acetic acid to propionate
[2]TL, TR, and NP were cellulose originating from *Trichoderma longibrachiatum* (for TL), *Trichoderma reesei* (for TR), or *Neocallimastix patriciarum* (for NP)
[3]Dose rate as IU/g of DM forage substrate; Mean = mean for individual EC across dose rates except dose rate of 0; 0 = control without added EC
[4]SEM for EC × DR
[5]EC = exogenous cellulase; DR = dose rate; L = linear effect of DR; Q = quadratic effect of DR; NS means $P > 0.10$; EC × DR = interaction between EC and DR
[6]SEM for pooled mean of EC excluding the dose rate of 0

The concentration of NH_3-N in EC treatments was higher than 5 mg/dL, indicating the growth rate of bacteria would not be restricted. On the other hand, NH_3-N is the main source of nitrogen in the synthesis of rumen bacteria, and 18% to 100% of bacterial nitrogen originates from NH_3-N (Salter et al. 1979). The treatments of EC or TR and NP had lower NH_3-N concentrations than control or TL, implying that forage supplemented EC or EC of TR and NP could enhance the utilization of nitrogen by bacteria.

Changes in TVFA and individual volatile fatty acid (VFA) corresponded to increased fiber degradation of forage. Addition of EC increased the concentration of TVFA of forage, and this was consistent with the findings of previous studies (Eun and Beauchemin 2008; Giraldo et al. 2008). These indicated that the activity of bacteria in degrading fiber had been promoted by EC, especially for EC of NP. The ratio of acetate to propionate of EC treatment was not decreased, or even increased compared with control, which was inconsistent with the findings of previous studies (Eun and Beauchemin 2008; Giraldo et al. 2008). Eun and Beauchemin (2007) found that the addition of single or combined enzyme numerically decreased the ratio of acetate to propionate of alfalfa. The inconsistent of enzyme addition on the changes of VFA composition may relate to the enzyme activities added, the forage substrates, diet of donor animal and donor species used. The ratio of acetate to propionate was lower for forage supplemented with EC at 37 IU/g of DM than that of higher dosage indicating that it is beneficial for diets supplementing EC at lower dosage from increasing availability of glucogenic precursors to ruminants.

Effects of Exogenous Xylanase Addition For MS, there was no effect of EPX on individual VFAs or NH_3-N concentration, except for a trend effect of EPX on NH_3-N concentration ($P = 0.061$) after 48 h (Table 12). Increasing EPX dose increased acetate proportion (linear effect, $P < 0.01$; quadratic effect, $P < 0.05$) at 24 h, but not at 48 h. The TVFA was increased (linear effect, $P < 0.01$) with increasing dose at 48 h.

For RS, there was no effect of EPX on individual VFAs or NH_3-N concentration (Table 13). Increasing EPX dose increased (linear effect, $P < 0.01$) TVFA at 24 h, but not at 48 h. Butyrate proportion and NH_3-N concentration were affected (linear effect, $P < 0.05$) by dose at 48 h.

For GM, there was no effect of EPX on individual VFAs or NH_3-N concentration (Table 14). Increasing EPX dose affected acetate (linear and quadratic effect, $P < 0.05$, 24 h; linear effect, $P < 0.05$, 48 h) and butyrate (linear and quadratic effect, $P < 0.01$, 24 h; linear effect, $P < 0.01$, 48 h) concentrations at 24 h and 48 h, TVFA (linear effect, $P = 0.062$; quadratic effect, $P < 0.05$) only at 24 h and NH_3-N concentration (quadratic effect, $P < 0.05$) only at 48 h.

The greater CH_4 production with EPX with RS and GM, but not MS, also indicated an enzyme-forage specificity. The greater CH_4 production with EPX might have resulted from greater acetate production. Acetate production is associated with the release of H_2, which is utilized by methanogens to form CH_4 (Stewart et al. 1997). Similarly, several studies have reported that fibrolytic enzymes increased

Table 12 VFA profiles and NH_3-N concentrations of in vitro incubation medium for maize stover[a] (from He et al. 2015)

Item		24 h of incubation						48 h of incubation					
		Acetate	Propionate	Butyrate	A:P	TVFA	NH_3-N	Acetate	Propionate	Butyrate	A:P	TVFA	NH_3-N
EPX group[b]	EPX1	63.9	19.2	7.3	3.43	48.7	87.5	65.0	20.9	6.5	3.12	70.1	97.9
	EPX2	65.5	18.0	7.2	3.76	46.0	87.9	65.0	21.2	6.5	3.07	76.2	89.2
	EPX3	64.1	18.2	7.7	3.63	44.0	78.6	64.9	21.2	6.6	3.07	75.6	92.3
	EPX4	63.0	17.6	7.7	4.13	42.8	81.7	64.7	21.2	6.7	3.06	76.6	97.8
	EPX5	64.0	19.7	6.8	3.33	47.2	89.4	64.3	21.4	6.8	3.02	76.7	90.1
$S.E.M_{EPX}$		1.42	0.97	0.48	0.322	4.07	3.74	0.41	0.20	0.13	0.051	2.41	3.56
Dose[c]	0	61.0	17.8	7.7	3.61	45.4	86.9	65.1	20.8	6.5	3.14	70.9	97.0
	20	64.6	18.2	7.4	3.98	45.4	86.2	64.3	21.2	6.8	3.03	72.1	90.3
	50	65.6	19.0	6.9	3.52	45.3	83.4	64.9	21.2	6.6	3.07	79.8	91.1
	80	65.1	19.1	7.4	3.51	46.8	83.8	64.8	21.3	6.6	3.04	77.6	95.5
$S.E.M_{Dose}$		1.33	0.87	0.44	0.289	3.64	3.36	0.37	0.18	0.11	0.041	2.16	3.32
P-value[d]													
EPX		NS	NS	NS	NS	NS	NS	NS	NS	NS	NS	NS	0.06
Dose		<0.01	NS	NS	NS	NS	NS	NS	NS	NS	NS	0.01	0.03
$Dose_{Linear}$		<0.01	NS	NS	NS	NS	NS	NS	0.10	NS	NS	<0.01	NS
$Dose_{Quadratic}$		0.04	NS	NS	NS	NS	NS	NS	NS	NS	NS	NS	0.03
EPX × Dose		NS	NS	NS	NS	NS	NS	NS	NS	NS	NS	NS	NS

[a]TVFA expressed as mM and individual VFA expressed as moles/100 moles; NH_3-N expressed as mg/dL; TVFA = total volatile fatty acids; TVFA = total VFA; NH_3-N = ammonia nitrogen; EPX = enzyme product with high xylanase activity

[b]EPX1, xylanase gene of which was derived from a strain of *Trichoderma harzianum*; EPX2, xylanase gene of which was derived from a strain of *Trichoderma reesei*; EPX3, xylanase gene of which was derived from a strain of *Orpinomyces*; EPX4, xylanase gene of which was derived from a strain of *Aspergillus oryzae*; EPX5 was produced from *Trichoderma longibrachiatum*

[c] Dose as units of xylanase/g of DM forage substrate; one enzymatic unit was expressed as μmol xylose/min/g or mL enzyme from oat spelt xylan, using a standard curve made with xylose (from 0 to 0.8 mg/mL) obtained under identical conditions

[d]NS means $P > 0.10$; $Dose_{Linear}$ = linear effect of dose; $Dose_{Quadratic}$ = quadratic effect of dose

Table 13 Total VFA profiles and NH_3-N concentrations of in vitro incubation medium for rice stover[a] (from He et al. 2015)

Item		24 h of incubation						48 h of incubation					
		Acetate	Propionate	Butyrate	A:P	TVFA	NH_3-N	Acetate	Propionate	Butyrate	A:P	TVFA	NH_3-N
EPX group[b]	EPX1	63.9	22.1	5.1	2.93	41.6	55.9	64.5	21.9	5.5	2.96	59.7	84.0
	EPX2	63.9	22.3	5.1	2.90	43.9	55.7	65.1	21.7	5.3	3.02	57.4	85.5
	EPX3	63.7	22.3	5.1	2.91	41.2	69.4	64.9	21.8	5.4	3.00	59.1	89.0
	EPX4	63.9	22.4	5.1	2.89	40.7	62.1	64.7	22.1	5.4	2.95	61.8	91.0
	EPX5	64.0	22.2	5.1	2.92	42.5	55.2	65.3	21.6	5.4	3.04	62.9	87.8
$S.E.M_{EPX}$		0.73	0.63	0.07	0.115	2.13	7.31	0.70	0.45	0.09	0.091	3.36	5.44
Dose[c]	0	64.1	21.8	5.1	3.00	35.6	65.8	64.6	22.0	5.5	2.97	58.5	78.1
	20	63.8	22.4	5.1	2.88	44.5	63.1	64.7	21.8	5.4	2.99	60.4	89.9
	50	63.6	22.4	5.1	2.88	42.5	53.5	65.0	21.8	5.4	3.00	59.7	87.4
	80	64.1	22.5	5.0	2.88	45.4	56.2	65.3	21.8	5.3	3.02	62.0	94.3
$S.E.M_{Dose}$		0.66	0.56	0.06	0.103	1.94	6.54	0.62	0.40	0.08	0.081	3.03	4.88
P-value[d]													
EPX		NS	NS	NS	NS	NS	NS	NS	NS	NS	NS	NS	NS
Dose		NS	NS	NS	NS	<0.01	NS	NS	NS	NS	NS	NS	0.11
$Dose_{Linear}$		NS	NS	NS	NS	<0.01	NS	NS	NS	0.05	NS	NS	0.04
$Dose_{Quadratic}$		NS	NS	NS	NS	0.10	NS	NS	NS	NS	NS	NS	NS
EPX × Dose		NS	NS	NS	NS	NS	NS	NS	NS	NS	NS	NS	NS

[a]TVFA expressed as mM and individual VFA expressed as moles/100 moles; NH_3-N expressed as mg/dL; TVFA = total volatile fatty acids; TVFA = total VFA; NH_3-N = ammonia nitrogen; EPX = enzyme product with high xylanase activity

[b]EPX1, xylanase gene of which was derived from a strain of *Trichoderma harzianum*; EPX2, xylanase gene of which was derived from a strain of *Trichoderma reesei*; EPX3, xylanase gene of which was derived from a strain of *Orpinomyces*; EPX4, xylanase gene of which was derived from a strain of *Aspergillus oryzae*; EPX5 was produced from *Trichoderma longibrachiatum*

[c]Dose as units of xylanase/g of DM forage substrate; one enzymatic unit was expressed as μmol xylose/min/g or mL enzyme from oat spelt xylan, using a standard curve made with xylose (from 0 to 0.8 mg/mL) obtained under identical conditions

[d]NS means $P > 0.10$; $Dose_{Linear}$ = linear effect of dose; $Dose_{Quadratic}$ = quadratic effect of dose

Table 14 Total VFA profiles and NH_3-N concentrations of in vitro incubation medium for Guimu No. 1 grass[a] (from He et al. 2015)

Item		24 h of incubation						48 h of incubation					
		Acetate	Propionate	Butyrate	A:P	TVFA	NH_3-N	Acetate	Propionate	Butyrate	A:P	TVFA	NH_3-N
EPX group[b]	EPX1	66.8	20.7	5.0	3.23	48.0	87.0	67.1	20.6	5.4	3.27	71.9	97.4
	EPX2	67.4	20.7	4.9	3.28	48.8	87.2	67.1	20.7	5.3	3.25	69.5	89.0
	EPX3	67.4	20.6	4.9	3.28	47.7	78.1	67.5	20.5	5.2	3.29	71.2	92.0
	EPX4	66.4	20.9	5.1	3.18	52.7	81.1	67.0	20.8	5.4	3.23	75.4	97.6
	EPX5	67.3	20.5	4.9	3.29	51.2	89.0	66.9	20.6	5.5	3.26	72.3	89.7
$S.E.M_{EPX}$		0.38	0.29	0.08	0.059	2.81	3.64	0.31	0.23	0.07	0.049	3.77	3.14
Dose[c]	0	65.9	20.8	5.2	3.17	43.3	86.2	67.8	20.3	5.2	3.34	72.9	96.8
	20	67.8	20.6	4.8	3.30	50.9	85.8	66.9	20.7	5.4	3.24	68.9	90.1
	50	67.1	20.8	4.9	3.24	54.5	82.5	67.1	20.7	5.4	3.26	72.7	90.8
	80	67.5	20.6	4.9	3.30	50.1	83.3	66.6	20.9	5.5	3.20	73.7	94.7
$S.E.M_{Dose}$		0.35	0.26	0.08	0.052	2.51	3.51	0.28	0.21	0.06	0.044	3.49	2.81
P-value[d]													
EPX		NS	NS	0.10	NS	NS	NS	NS	NS	0.14	NS	NS	0.06
Dose		<0.01	NS	<0.01	NS	0.02	NS	0.03	NS	<0.01	0.12	NS	0.05
$Dose_{Linear}$		0.01	NS	<0.01	NS	0.06	NS	0.01	0.08	<0.01	0.04	NS	NS
$Dose_{Quadratic}$		0.04	NS	<0.01	NS	0.01	NS	NS	NS	NS	NS	NS	0.02
EPX × dose		NS	NS	NS	NS	NS	NS	NS	NS	NS	NS	NS	NS

[a]TVFA expressed as mM and individual VFA expressed as moles/100 moles; NH_3-N expressed as mg/dL; TVFA = total volatile fatty acids; TVFA = total VFA; NH_3-N = ammonia nitrogen; EPX = enzyme product with high xylanase activity

[b]EPX1, xylanase gene of which was derived from a strain of *Trichoderma harzianum*; EPX2, xylanase gene of which was derived from a strain of *Trichoderma reesei*; EPX3, xylanase gene of which was derived from a strain of *Orpinomyces*; EPX4, xylanase gene of which was derived from a strain of *Aspergillus oryzae*; EPX5 was produced from *Trichoderma longibrachiatum*

[c]Dose as units of xylanase/g of DM forage substrate; one enzymatic unit was expressed as μmol xylose/min/g or mL enzyme from oat spelt xylan, using a standard curve made with xylose (from 0 to 0.8 mg/mL) obtained under identical conditions

[d]NS means $P > 0.10$; $Dose_{Linear}$ = linear effect of dose; $Dose_{Quadratic}$ = quadratic effect of dose

in vitro CH_4 production in grass hay alone or with concentrates (Dong et al. 1999; Giraldo et al. 2007, 2008). However, the lack of effect of EPX addition on CH_4/IVDMD production for MS and RS indicated that the use of EPX with these two forages did not increase the intensity of CH_4 production, and thus any increase in CH_4 production with added EPX was due to an increase in forage degradation (i.e., IVDMD).

Given the substantial effects of EPX on forage degradation, it was considered important to evaluate changes in the fermentation end products during incubation, with VFA and NH_3-N concentrations as indicators. It was expected that increased VFA and NH_3-N production would accompany any improvement of forage degradability. However, in the present study, the unchanged individual VFAs and NH_3-N concentrations for RS using EPX were not expected, as the degradation was enhanced with increasing EPX. The lack of complete consistency between degradation and fermentation end products production might be due to the type of substrates, EPX sources, and the efficiency of enzymes (i.e., ruminal enzyme and EPX) on microbial growth, which affect forage fiber utilization.

2.3 Conclusion

Exogenous cellulase originating from TR and NP improved in vitro DMD of forage, decreased pH value, and NH_3-N and CH_4 production of forage compared to TL. Adding EC of NP in the forage had higher TVFA, acetate, propionate, and the ratio of acetate to propionate compared with TL and TR. Feeding EC with forage could improve DMD, NDFD, and TVFA, and decrease pH value, NH_3-N and CH_4 production, and their optimum DR varied mainly depending on the forage types. In general, low DR of EC resulted in increases in DMD and NDFD and decreases in CH_4 and NH_3-N compared with medium and higher DR. It is recommended that the treatments of TR and NP be further evaluated in animal feeding studies.

Although responses to EPX were varied among the forages evaluated, the present results showed that several EPX substantially increased the in vitro ruminal degradation of MS, RS, and GM, but responses were forage-specific. The current results indicate that EPX has the potential to improve ruminal digestion of local low-quality forages in China. The source of organism used to derive the EPX gene might also be a consideration in the effects of EPX on forage degradation, with the evidence of more positive effects on degradability and initial fractional rate of degradation from RS when applied an EPX derived from *Aspergillus oryzae*.

3 In Vivo Evaluation of Exogenous Fibrolytic Enzyme in Goat And Cattle Feeding System

3.1 *Materials and Methods*

3.1.1 Evaluation of Candidate Fibrolytic Enzymes in the Feeding System of Goats

Five Xiangdong female black goats (20.2 ± 1.5 kg, a local breed in the south of China), fitted with permanent ruminal plastic cannulae, were used in a 5 × 5 Latin square design. A 2 × 2 factorial arrangement of treatments [two exogenous enzymes (cellulase and xylanase) and two dose rates] and another treatment without exogenous enzyme addition as control were used in this study. The cellulase (powder form with an activity of 243 U/g) derived from *Neocallimastix patriciarum* was offered at the rates of 0.8 and 1.6 g/kg of DM intake (namely CEL0.8 and CEL1.6, respectively). The xylanase (liquid form with an activity of 31,457 U/mL) derived from *Aspergillus oryzae, which* was fed at the rates of 1.4 and 2.2 mL/kg of DM intake (namely XYL1.4 and XYL2.2, respectively). The diet (rice straw, 55%; concentrate, 45%) was formulated to meet 1.3 times maintenance metabolizable energy (ME) requirements. The ingredients and chemical composition of the diet are shown in Table 15. The cellulase was equably mixed with concentrate, and the xylanase was sprayed onto the rice straw before feeding. The digestion of OM, NDF and ADF, the metabolism of nitrogen, and the emission of methane were measured. The rumen fermentation characteristics were estimated.

Table 15 Dietary ingredients and chemical composition of the goats (DM basis) (Lu et al. 2015)

Ingredient, %		Chemical composition, %[b]	
Rice straw	55.0	ME (MJ/kg)	8.86
Ground corn grain	31.1	OM	89.3
Soybean meal	9.90	CP	9.33
Fat powder	2.14	NDF	48.2
Salt	0.86	ADF	24.0
Premix[a]	1.0	Ca	0.24
		P	0.17

[a]Premix contained per kilogram: 227 g $MgSO_4 \cdot H_2O$, 12.5 g $FeSO_4 \cdot 7H_2O$, 2.8 g $CuSO_4 \cdot 5H_2O$, 12.2 g $MnSO_4 \cdot H_2O$, 13.4 g $ZnSO_4 \cdot H_2O$, 20 mg Na_2SeO_3, 50 mg KI, 35 mg $CoCl_2 \cdot 6H_2O$, 90,000 IU vitamin A, 17,000 IU vitamin D, and 17,500 IU vitamin E

[b]ME = metabolizable energy, calculated by the data of Lu and Zhang (1996). OM = organic matter; CP = crude protein; NDF = neutral detergent fiber; ADF = acid detergent fiber; Ca = calcium; P = phosphorus

Table 16 Dietary ingredients and chemical components of the cattle

Ingredient (kg)	Control	Treat 1	Treat 2
Rice straw	45	45	45
Canola meal	9.5	9.5	9.5
Distillers grains	11	11	11
Corn	16.2	16.2	16.2
Soybean meal	10.0	10.0	10.0
Wheat bran	4.5	4.5	4.5
$CaCO_3$	0.75	0.75	0.75
Premix	2.0	2.0	2.0
Salt	0.60	0.60	0.60
Sodium bicarbonate	0.50	0.50	0.50
Cellulase(g)	0.00	60.00	180.00
Chemical composition, %			
ME(Mcal/kg)	2.14	2.14	2.14
CP	13.8	13.65	14.83
Ca	0.50	0.57	0.55
P	0.36	0.30	0.44

Premix contained per kilogram: 50 g Mg, 2.5 g Fe, 0.4 g Cu, 2 g Mn, 1.5 g Zn, 10 mg Se, 25 mg I, 5 mg Co, 80,000 IU vitamin A, 12,500 IU vitamin D, and 1250 mg vitamin E
ME = metabolizable energy, calculated value. OM = organic matter; CP = crude protein; NDF = neutral detergent fiber; ADF = acid detergent fiber; Ca = calcium; P = phosphorus

3.1.2 Evaluation of Candidate of Fibrolytic Enzymes Estimation in Cattle

Eighteen Hunan Black cattle (body weight 335 ± 43 kg) were used in this experiment. They were settled in three treatments, and each treatment had 6 cattle. The control group had no enzyme addition, and treatment one and treatment two had received cellulase extracted from *TL* at 60 and 180 g/100 kg diet, respectively. The diet compositions of the cattle are listed in Table 16. The digestion of DM, CP, NDF, and ADF, methane production, and the rumen fermentation function were estimated. Also, the plasma physiology and biochemistry parameters were measured.

3.2 *Results and Discussion*

3.2.1 In Goats

Apparent Digestibility of Nutrition The influence of exogenous cellulase and xylanase supplementation on nutrient digestibility of goats is shown in Table 17. Supplementation of cellulase and xylanase did not affect the intake of dry matter, organic matter, and fiber and their digestibility. In contrast, fecal excretion of dry matter and organic matter increased with increasing exogenous xylanase supplementation. Cellulase interacted with xylanase on fecal excretion of dry matter and

Table 17 Effects of dietary addition of cellulase (CEL[a]) and xylanase (XYL[b]) on nutrient intake, fecal output, and apparent total tract digestibility in goats (from Lu et al. 2015)

		Intake [g DM/day]			Fecal output [g DM/day]			Digestibility [%]		
		OM[c]	NDF[d]	ADF[e]	OM	NDF	ADF	OM	NDF	ADF
Treatments										
Control[f]		476	295	148	151	97.4	60.3	68.3	66.9	59.3
CEL0.8	XYL1.4	462	284	141	141	88.7	56.0	69.6	68.9	60.3
	XYL2.2	460	282	140	155	96.9	61.3	66.5	65.9	56.7
CEL1.6	XYL1.4	467	288	144	150	95.3	60.1	68.0	67.0	58.3
	XYL2.2	477	296	148	154	97.4	63.5	67.6	67.1	57.3
SEM[g]		8.38	6.85	4.22	6.78	5.30	3.22	1.17	1.40	1.46
p-values										
CEL		NS	NS	NS	NS	NS	NS	NS	NS	NS
XYL		NS	NS	NS	NS	NS	NS	NS	NS	NS
CEL × XYL		NS	NS	NS	NS	NS	NS	NS	NS	NS
Contrasts										
CEL vs control		NS	NS	NS	NS	NS	NS	NS	NS	NS
XYL vs control		NS	NS	NS	NS	NS	NS	NS	NS	NS

Notes:
[a]CEL0.8 and CEL1.6, Cellulase added at 0.8 and 1.6 g/kg DM (rice straw), respectively
[b]XYL1.4 and XYL2.2, Xylanase added at 1.4 and 2.2 mL/kg DM (rice straw), respectively
[c]OM, Organic matter
[d]NDF, Neutral detergent fiber
[e]ADF, Acid detergent fiber
[f]Without enzyme addition
[g]SEM, Standard error of means; NS means $P > 0.10$

organic matter. Feces excretion of dry matter and organic matter was increased when cellulase was supplemented at 0.8 g/kg DM with xylanase and decreased when cellulase was supplemented at 1.6 g/kg DM with xylanase. There are no significant differences between control and enzyme supplementation treatment in dry matter, organic matter and fiber intake, and their digestibility.

More than 0.23 MJ of gross energy was ingested by the animals for the diet supplemented cellulase at 1.6g/kg DM than at 0.8g/kg DM (Table 18). More than 184.5 kJ of gross energy was excreted for diet supplementing xylanase at 2.2 g/kg DM than at 1.4 g/kg DM. Interaction effect on fecal energy excretion was noted between cellulase and xylanase. Fecal energy was increased when cellulase was supplemented at 0.8 g/kg DM with xylanase and decreased when cellulase was supplemented at 1.6 g/kg DM with xylanase. Methane emission and energy digestion were not affected by enzyme supplementation. Compared with control, the methane emission numerically decreased, but energy utilization increased for diet supplement cellulase at 0.8 g/kg DM and xylanase supplemented at 1.4 g/kg DM.

No effects were observed in crude protein intake, excretion, and digestion in response to enzyme supplementation (Table 19). While for feces crude protein

Table 18 Effects of dietary cellulose (CEL[a]) and xylanase (XYL[b]) addition on energy balance in goats (from Lu et al. 2015)

		Ingested gross energy [MJ/day]	Fecal energy [MJ/day]	Urinary energy [MJ/day]	CH_4 energy [MJ/day]	Digestible energy [MJ/kg DM]	Metabolizable energy [MJ/kg DM]
Treatment							
Control[c]		9.33	3.04	0.22	0.74	15.57	13.19
CEL0.8	XYL1.4	9.03	2.79	0.17	0.71	16.67	14.30
	XYL2.2	9.00	3.06	0.28	0.66	15.58	13.11
CEL1.6	XYL1.4	9.15	2.96	0.19	0.71	15.87	13.56
	XYL2.2	9.34	3.06	0.21	0.72	15.53	13.24
SEM[d]		0.164	0.134	0.027	0.04	0.552	0.508
p-values							
CEL		NS	NS	NS	NS	NS	NS
XYL		NS	NS	0.06	NS	NS	NS
CEL × XYL		NS	NS	NS	NS	NS	NS
Contrasts							
CEL vs control		NS	NS	NS	NS	NS	NS
XYL vs control		NS	NS	NS	NS	NS	NS

Notes:
[a]CEL0.8 and CEL1.6, Cellulase added at 0.8 and 1.6 g/kg DM (rice straw), respectively
[b]XYL1.4 and XYL2.2, Xylanase added at 1.4 and 2.2 mL/kg DM (rice straw), respectively
[c]Without enzyme addition
[d]SEM, Standard error of means; NS means $P > 0.10$

excretion and nitrogen retention, interaction effects were noted between cellulase and xylanase. No significant differences were noted in crude protein intake, excretion, digestion, and N retention between the control and enzyme supplemented groups. Compared with control, the crude protein digestibility and nitrogen retention were improved by 7.9% and 7.1% for the diet supplemented with cellulase at 0.8 g/kg DM and xylanase at 1.4 g/kg DM, respectively.

Compared with control, exogenous cellulase and xylanase supplementation had no positive effects on apparent total tract digestibility of nutrients and energy utilization (Tables 17 and 18). Similarly, Lewis et al. (1999) have demonstrated that the digestibility of nutrients was not improved by feeding fibrolytic enzymes in dairy cows, while Bhasker et al. (2013a, b) have also affirmed that supplementing exogenous fibrolytic enzymes did not alter the nutrient utilization in sheep fed on a maize stover-based total mixed ration. On the contrary, other studies proved that dietary inclusions of fibrolytic enzymes could increase nutrient digestibility to a certain extent in ruminants (Gaafar et al. 2010). These discrepancies may be due to differences in experimental animals (cows, sheep, or lactating buffaloes). For example, Yang et al. (2000) have demonstrated that the digestibility of diet was improved by exogenous enzyme supplementation when evaluated in dairy cows, but not in sheep. Furthermore, the differences might be related to the dose of commercial enzymes. As for N balance, it was noted that digested N, retained N, and its

Table 19 Effects of dietary cellulase (CEL[a]) and xylanase (XYL[b]) addition on N balance in goats (from Lu et al. 2015)

		Ingested N [g/day]	Fecal N [g/day]	Urinary N [g/day]	Digested N (DN) [g/day]	Retained N (RN) [g/day]	Availability [% of RN]	Availability [% of DN]	Digestibility [%]
Treatment									
Control[c]		9.01	2.49	2.41	6.52	4.11	44.1	60.5	71.9
CEL0.8	XYL1.4	8.85	2.10	1.91	6.75	4.85	53.8	70.7	76.5
	XYL2.2	8.83	2.33	3.35	6.50	3.15	34.6	46.3	73.4
CEL1.6	XYL1.4	9.47	2.65	1.63	6.82	5.64	57.5	77.1	71.5
	XYL2.2	8.65	2.36	2.67	6.29	3.62	42.9	58.2	73.4
SEM[d]		0.392	0.205	0.378	0.445	0.424	4.69	5.17	2.49
p-values									
CEL		NS	NS	NS	NS	NS	NS	NS	NS
XYL		NS	NS	0.014	NS	0.003	0.009	0.004	NS
CEL × XYL		NS	NS	NS	NS	NS	NS	NS	NS
Contrasts									
CEL vs control		NS	NS	NS	NS	NS	NS	NS	NS
XYL vs control		NS	NS	NS	NS	NS	NS	NS	NS

Notes:
[a]CEL0.8 and CEL1.6, Cellulase added at 0.8 and 1.6 g/kg DM (rice straw), respectively
[b]XYL1.4 and XYL2.2, Xylanase added at 1.4 and 2.2 mL/kg DM (rice straw), respectively
[c]Without enzyme addition
[d]SEM, Standard error of means; NS means $P > 0.10$

availability were increased at the low dose of xylanase supplementation (1.4 mL/kg DMI) due to the reduction of urinary N output (Table 19). Similarly, Shekhar et al. (2010) reported that dietary inclusion of cellulase and xylanase mixture at a low dose could improve nutrient digestibility in Murrah buffaloes receiving a total mixed ration containing wheat straw (45%), green maize (15%), and concentrate (40%).

Rumen Fermentation From Table 20, we can find that there was no significant difference in rumen fermentation parameters between different dosages of cellulase or xylanase. No significant differences were also observed between control and enzyme supplementation treatment. These results mean that the diet supplemented with exogenous cellulase and xylanase at these dosages did not affect rumen fermentation profiles. Compared with the control treatment, the mole percent proportion of propionate and mole concentration of total VFA numerically increased by 2.8% and 6.1%, but the ratio of acetate to propionate numerically decreased by 18.1%.

Although the ruminal pH values were decreased by cellulase and xylanase addition in the diets of goats, it still ranged within the normal physiological threshold (Russi et al. 2002). The ruminal NH_3-N and total VFA concentration, individual VFA molar proportion, and ratio of acetate to propionate were not altered in response to cellulose and xylanase addition in the diets of goats, except that the molar proportion of ruminal butyrate showed an incremental tendency with the higher dose of xylanase (Table 20). Likewise, previous studies have already reported the lack of response to exogenous fibrolytic enzyme addition, and this has been reported for ruminal fermentation patterns of ruminants (Peters et al. 2010; Hernández et al. 2011; Chung et al. 2012). In contrast, some studies suggested that enzyme application increased total VFA concentration (Gaafar et al. 2010) and the molar proportion of propionate, but decreased the molar proportion of butyrate and acetate to propionate ratio (Tirado-Estrada et al. 2011) and NH_3-N concentration (Bhasker et al. 2013a, b) in the rumen. The inconsistency in ruminal fermentation parameters resulting from dietary enzyme addition would be ascribed to above-mentioned factors, which were pointed out by Beauchemin et al. (2004a, b).

Enzyme Activities of Rumen Effects of enzyme supplementation on rumen enzyme activities are shown in Table 21. Extracellular and intracellular carboxymethyl cellulase and xylanase activities were not affected by the dosage of cellulase and xylanase supplementation. Cellulase did not interact with xylanase on enzyme activities. Compared with the control treatment, extracellular xylanase and total xylanase activities in rumen decreased by 28% and 22.5%, respectively, in response to exogenous cellulase and xylanase supplementation.

Even though the supplementation of exogenous cellulase and xylanase at these levels did not profoundly improve nutrient utilization and decrease the emission of methane in goats, the nutrient utilization and methane emission could be numerically decreased by the supplementation of cellulase and xylanase at 0.8 g/kg DM and 1.4 g/kg DM, respectively. It may suggest that the exogenous enzyme

Table 20 Effects of dietary cellulose (CEL[a]) and xylanase (XYL[b]) addition on the ruminal pH, NH_3-N, and molar proportions of volatile fatty acids (VFAs) in goats (from Lu et al. 2015)

		pH	NH_3-N [mg/dL]	TVFA [mM]	Acetate [mol/100 mol]	Propionate [mol/100 mol]	Butyrate [mol/100 mol]	Ratio A:P
Treatment								
Control[c]		6.38	8.83	66.2	72.1	16.0	8.76	4.53
CEL0.8	XYL1.4	6.27	9.51	74.2	72.5	16.3	8.01	4.63
	XYL2.2	6.19	9.74	74.3	70.2	17.6	9.33	4.11
CEL1.6	XYL1.4	6.23	9.15	75.8	72.9	16.8	7.77	4.37
	XYL2.2	6.19	9.04	66.5	69.7	17.2	9.84	4.17
SEM[d]		0.027	0.642	3.98	2.23	1.23	0.779	0.424
p-values								
CEL		NS	NS	NS	NS	NS	NS	NS
XYL		0.028	NS	NS	NS	NS	NS	NS
CEL × XYL		NS	NS	NS	NS	NS	NS	NS
Contrasts								
CEL vs control		<0.001	NS	NS	NS	NS	NS	NS
XYL vs control		<0.001	NS	NS	NS	NS	NS	NS

Notes
[a]CEL0.8 and CEL1.6, Cellulase added at 0.8 and 1.6 g/kg DM (rice straw), respectively
[b]XYL1.4 and XYL2.2, Xylanase added at 1.4 and 2.2 mL/kg DM (rice straw), respectively
[c]Without enzyme addition
[d]SEM, Standard error of means; NS means $P > 0.10$

Table 21 Effect of dietary cellulase (CEL[a]) and xylanase (XYL[b]) addition on the ruminal enzyme activity in goats. (from Lu et al. 2015)

		Carboxymethyl cellulase activity [IU/mL]			Xylanase activity [IU/mL]		
		Extracellular	Intracellular	Total	Extracellular	Intracellular	Total
Treatment							
Control[c]		0.10	0.06	0.16	0.50	0.35	0.85
CEL0.8	XYL1.4	0.08	0.05	0.12	0.50	0.34	0.83
	XYL2.2	0.08	0.05	0.13	0.44	0.33	0.77
CEL1.6	XYL1.4	0.08	0.06	0.15	0.53	0.37	0.90
	XYL2.2	0.08	0.06	0.14	0.41	0.33	0.74
SEM[d]		0.014	0.016	0.03	0.053	0.031	0.07
p-values							
CEL		NS	NS	NS	NS	NS	NS
XYL		NS	NS	NS	NS	NS	NS
CEL × XYL		NS	NS	NS	NS	NS	NS
Contrasts							
CEL vs control		NS	NS	NS	NS	NS	NS
XYL vs control		NS	NS	NS	NS	NS	NS

Notes:
[a]CEL0.8 and CEL1.6, Cellulase added at 0.8 and 1.6 g/kg DM (rice straw), respectively
[b]XYL1.4 and XYL2.2, Xylanase added at 1.4 and 2.2 mL/kg DM (rice straw), respectively
[c]Without enzyme addition
[d]SEM, Standard error of means; NS means $P > 0.10$

supplementation shows the potential to improve nutrient utilization and decrease greenhouse gas emissions by ruminants.

In the current study, dietary enzyme application did not influence the activities of ruminal extracellular and intracellular CMCase and xylanase in goats and this might be the primary reason that the supplemented enzymes did not alter most measured parameters (Table 21). There are some evidence to prove that a direct administration of exogenous fibrolytic enzymes into the rumen leads to an increase in ruminal enzymatic activity (i.e., CMCase and xylanase) (Giraldo et al. 2008). The authors stated that the increase in the ruminal enzymatic activity was not exclusively due to a direct effect of the exogenous enzyme supplementation, which could stimulate the ruminal fibrolytic activity. These discrepancies may be due to differences in experimental animals (goats vs sheep), enzyme source, and activity (*N. patriciarum* vs. *Aspergillus niger* and *T. longibrachiatum*), application rate and mode (fed with concentrates vs delivered directly into the rumen), time of enzyme delivery, ruminal activity and stability of enzymes, enzyme-feed specificity (Beauchemin et al. 2004b).

Methane Emission Furthermore, dietary cellulase and xylanase addition did not influence the enteric methane emissions in goats, even when the methane emission data were corrected to DMI, OM intake, digested DM, and ingested GE (Table 22). Nevertheless, compared with the control, the addition of xylanase decreased the

Table 22 Effects of dietary cellulase (CEL[a]) and xylanase (XYL[b]) addition on the enteric methane emission in goats (from Lu et al. 2015)

		CH_4 emission				
		[L/kg DM intake]	[L/kg OM[c] intake]	[L/kg NDF[d] intake]	[L/kg ADF[e] intake]	CH_4 energy [% of ingested GE[f]]
Treatment						
Control[g]		42.8	46.5	77.5	160.0	8.0
CEL0.8	XYL1.4	39.5	43.0	70.4	142.3	7.9
	XYL2.2	35.7	38.9	62.6	125.5	7.4
CEL1.6	XYL1.4	39.5	43.0	70.6	141.7	7.8
	XYL2.2	39.4	42.9	69.5	139.7	7.7
SEM[h]		2.38	2.59	4.59	9.64	0.51
p-values						
CEL		NS	NS	NS	NS	NS
XYL		NS	NS	NS	NS	NS
CEL × XYL		NS	NS	NS	NS	NS
Contrasts						
CEL vs control		NS	NS	0.076	0.051	NS
XYL vs control		NS	NS	0.072	0.047	NS

Notes:
[a]CEL0.8 and CEL1.6, Cellulase added at 0.8 and 1.6 g/kgDM (rice straw), respectively
[b]XYL1.4 andXYL2.2, Xylanase added at 1.4 and 2.2 mL/kg DM (rice straw), respectively
[c]OM, Organic matter
[d]NDF, Neutral detergent fiber
[e]ADF, Acid detergent fiber
[f]Without enzyme addition
[g]GE, Gross energy
[h]SEM, Standard error of means; NS means $P > 0.10$

methane emission when corrected to ADF intake ($p = 0.047$) and the inclusion of cellulase tended to decline this parameter. Similarly, the methane emission corrected to NDF intake tended to decrease in response to cellulase and xylanase supplementation.

It was also observed that dietary cellulase and xylanase addition did not influence the enteric methane emissions expressed as daily emissions corrected to DMI, OMI, digested DM, and % of ingested GE, respectively (Table 22). Interestingly, the enteric CH_4 emissions, corrected to ADF intake, were decreased after supplementation of xylanase and tended to be decreased by cellulose. Meanwhile, methane emission corrected to NDF intake tended to decline by cellulase and xylanase in comparison with the control. These might imply that the enteric methane emissions are more closely related to dietary ADF content. In recent years, the effects of enzyme application in the diets of ruminants on enteric methane emissions have been equivocal. For example, Arriola et al. (2011) reported that enteric methane emissions could be significantly decreased by dietary fibrolytic enzyme addition in lactating dairy cows, mainly when diets are rich in concentrates fed. In that study, the authors suggested that a dietary supplementation of fibrolytic enzymes contributes to a reduction of greenhouse gas production from dairy cows. However, Chung et al. (2012) have pointed out that the high doses of applicated enzymes could produce more CH_4 deriving from lactating cows. From our point of view, this controversy in the effect of enzyme addition on enteric methane emission can be due to the difference in enzyme source and activity, application rate and mode, and enzyme-feed specificity (Beauchemin et al. 2004a, b). In addition, our previous study indicated that supplementation of cellulase (derived from N. patriciarum, same as the present study) declines methane production in vitro (Tang et al. 2013). However, data on screening exogenous enzymes in vitro was insufficient to reflect the effectiveness of enzyme application in vivo. Moreover, the unchanged enteric methane emission in response to dietary cellulase and xylanase addition was also reflected by the estimation of ruminal methanogens community diversity. In this study, the ruminal methanogens community (both copy numbers and diversity) was not influenced by enzyme addition, which was in agreement with the findings of Zhou et al. (2011). They suggest that the ruminal methanogens community is kept relatively stable for dairy cows fed on diets supplied with exogenous fibrolytic enzymes.

3.2.2 In Cattle

Rumen Fermentation Table 23 showed the rumen fermentation of cattle as cellulase supplemented in concentration. The concentration of NH_3-N and volatile fatty acid was not affected by the supplementation of cellulase. Also, the proportion of individual volatile fatty acids was not affected by cellulase supplementation, except for isobutyrate and isovalerate, which was the lowest as cellulose supplemented at the dose of 60 g/100 kg diet. There was no significant difference among the three treatments in the ratio of acetate to propionate.

Table 23 Effects of cellulase supplementation on NH_3-N concentration and volatile fatty acid production and proportion

	Control	Treat 1	Treat 2	SEM	*P* value
NH_3-N (mg/dL)	7.49	9.14	5.79	1.94	NS
Total VFA (mM)	101.7	92.6	109.9	16.12	NS
Acetate (%)	67.7	56.3	67.3	5.45	NS
Propionate (%)	18.8	15.8	18.9	2.16	NS
Butyrate (%)	12.4	10.9	13.1	1.63	NS
Isobutyrate (%)	1.63^a	1.15^b	1.52^a	0.102	0.004
Valerate (%)	1.44	1.18	1.45	0.226	NS
Isovalerate (%)	1.18^a	0.85^b	1.14^a	0.090	0.011
Acetate:Propionate	3.82	3.78	3.79	0.114	NS

Note: Control, Treat 1, and Treat 2 mean enzyme supplementation in diet were 0, 60, and 180 g/100 kg; NS means $P > 0.10$; ^{a,b} Means within a row for dose rates of 0 to 180 g/100 kg diet that do not have a common superscript differ at $P < 0.05$

Table 24 Effects of cellulase supplementation on nutrient digestibility

	Control	Treat 1	Treat 2	SEM	*P* value
DM					
Intake(kg/day)	7.67	7.50	7.21	0.651	NS
Excretion in fecal (kg/day)	3.26	2.91	2.86	0.152	NS
Digestibility (%)	56.0	60.3	59.3	3.07	NS
CP					
Intake(kg/day)	1.06	1.02	1.07	0.163	NS
Excretion in fecal (kg/day)	0.314[a]	0.270[b]	0.260[b]	0.0146	0.044
Digestibility (%)	67.0	70.7	72.7	4.09	NS
NDF					
Intake(kg/day)	3.11	3.18	2.93	0.144	NS
Excretion in fecal (kg/day)	0.732	0.770	0.788	0.0726	NS
Digestibility(%)	75.9	75.8	73.4	2.28	NS
ADF					
Intake (kg/day)	1.43	1.36	1.29	0.089	NS
Excretion in fecal (kg/day)	0.625	0.645	0.575	0.0593	NS
Digestibility (%)	54.7	51.4	54.0	6.09	NS

Note: Control, Treat 1, and Treat 2 mean enzyme supplementation in diet were 0, 60, and 180 g/100 kg; NS means $P > 0.10$; [a, b]Means within a row for dose rates of 0 to 180 g/100 kg diet that do not have a common superscript differ at $P < 0.05$

Rumen concentration of NH_3-N and total VFA, molar proportion of individual VFA, and ratio of acetate to propionate were not altered in response to cellulase addition in the cattle diet, except that the molar proportion of isobutyrate and isovalerate decreased when the cellulase was supplemented at a lower dose. Similarly, the lack of response of enzyme addition to rumen fermentation has been reported in previous studies (Peters et al. 2010; Hernández et al. 2011; Chung et al. 2012; Lu et al. 2015). However, some researchers demonstrated that exogenous enzyme supplementation enhanced rumen total VFA concentration of buffaloes (Gaafar et al. 2010) and lambs (Bhasker et al. 2013a), and the molar proportion of propionate (Tirado-Estrada et al. 2011), while it decreased rumen concentration of NH_3-N of sheep (Bhasker et al. 2013b). These discrepancies may result from the differences in enzyme source and activity, application rate and providing method, enzyme-feed specificity, and diet composition (Beauchemin et al. 2004a, b).

Nutrient Digestion and Methane Emission There were no effects ($P > 0.05$) of feeding dietary cellulase supplementation on daily ingested DM, CP, NDF, and ADF among all treatments (Table 24). The excretion and digestibility of DM, CP, NDF and ADF were not influenced by cellulase supplementation, except for the excretion of CP was higher ($P < 0.05$) in control than that of Treat 1 and Treat 2.

Dietary supplementation with exogenous fibrolytic enzyme did not significantly influence nutrient intake and digestion, and energy intake except for the excretion of CP. Similar results were also reported in previous studies. Lewis et al. (1999)

Table 25 Effects of cellulose supplementation on methane emission

	Control	Treat 1	Treat 2	SEM	*P* value
GE intake (MJ/day)	125	122	117	10.8	NS
Methane emission (L/day)	200	260	231	20.4	NS
Methane emission (L/kg DMD)	49.2	59.3	56.6	6.09	NS
Methane emission (L/kg NDFD)	87.3	108.7	109.9	11.49	NS
Methane energy emission(MJ/day)	7.95	10.32	9.19	0.812	NS
Proportion of methane energy to GE (%)	6.54	8.56	8.00	0.634	0.098

Note: Control, Treat 1 and Treat 2 mean enzyme supplementation in diet were 0, 60 and 180 g/100 kg; NS means $P > 0.10$

showed that the nutrient digestibility of dairy cows was not improved by fibrolytic enzyme addition. Bhasker et al. (2013b) reported that the supplementation of exogenous fibrolytic enzyme had no significant effect on nutrient digestibility for lambs offered with maize stover based total mixed ration. Also, Lu et al. (2015) reported that the exogenous cellulase and xylanase addition in the diet of goats did not increase nutrient utilization. On the contrary, some studies have proved that diets involving exogenous enzymes improved nutrient utilization for dairy cows (Gaafar et al. 2010). These discrepancies are likely related to the animal differences used in these studies. For example, Yang et al. (2000) have revealed that dairy cows offered with the diet inclusion of exogenous enzymes lead to a higher DM digestibility compared with control, but this phenomenon was not observed in sheep.

In the present study, though the positive effects of exogenous enzymes added in the concentrate on nutrient utilization were not significant, the digestibility of DM numerically increased by 4.3% and 3.3%, and the digestibility of CP numerically increased by 3.7% and 5.7%, respectively, compared to control, and the CP excretion significantly decreased by enzyme addition. In our previous study (Lu et al. 2015), the urinary N output of goats was decreased as the low dose of xylanase was supplemented. Yang et al. (2000) reported that total tract digestibility of DM was higher for enzyme added in concentrate than for control diet and intermediate for enzyme added in TMR when measured in dairy cows. Bowman et al. (2002) and Rode et al. (1999) have found that adding enzyme to concentrate was effective. Also, Yang et al. (1999) observed that enzyme product applied to dry forage or both dry forage and concentrate increased digestibility of OM and NDF compared with control. In contrast, Phipps et al. (2000) reported no difference between adding an enzyme product to concentrate or TMR. Reasons for no response when enzymes are applied to the diet are unclear.

Gross energy intake, methane emission, methane energy emission, and proportion of methane energy to gross energy did not differ among the three treatments (Table 25). But the proportion of methane energy to gross energy of control treatment tended to be lower than that of treatments 1 and 2.

For CH_4 emission, though exogenous enzyme addition did not affect it's production, treatment 1 and 2 increased CH_4 emission by 60 and 31 L/day, or by 10.1 and 7.4 L/kg DMD, respectively, compared with the control. In the previous study,

Table 26 Effects of cellulose supplementation on plasma physiology and biochemical parameters

	Control	Treat 1	Treat 2	SEM	*P* value
ALT (U/L)	19.9	28.7	25.9	4.23	NS
GLU (mmol/L)	4.02	4.00	4.06	0.194	NS
LDH (U/L)	997	1292	869	176.0	NS
TC (mmol/L)	3.24	3.14	3.08	0.132	NS
TG (mmol/L)	0.235	0.228	0.232	0.0339	NS
TP (g/L)	69.0	70.7	70.6	2.12	NS
UREA (mmol/L)	6.66	6.06	6.24	0.498	NS
α-AMY (U/L)	49.4	50.0	37.1	4.34	0.081

Note: Control, Treat 1, and Treat 2 mean enzyme supplementation in diet were 0, 60, and 180 g/100 kg. ALT: Alanine aminotransferase, GLU: glucose, LDH: lactic dehydrogenase, TC: serum total cholesterol, TG: triglyceride, TP: total protein, α-AMY: α-Amylase; NS means $P > 0.10$

Chung et al. (2012) reported that the high doses of enzyme application resulted in more CH_4 deriving from lactating cows. Higher enzyme application decreased enteric CH_4 emission in our experiment, and it was incomparable with them. However, Lu et al. (2015) found that exogenous enzyme inclusion in the diet of goats did not influence the enteric CH_4 emission as expressed as total daily emission [L/day] or digested DM. But, the authors found that the enteric CH_4 emissions tended to decrease as cellulase was applied to the goat diet. Furthermore, Arriola et al. (2011) estimated a reduction in methane production based on a lower acetate to propionate ratio in ruminal fluid as dairy cows fed a fibrolytic enzyme. Beauchemin et al. (2008) proposed using in vitro studies that dietary enzyme feed additives may help mitigate enteric emissions based on fibrolytic enzymes, improve fiber degradation, and decrease the ratio of acetate to propionate in fermentation fluid (Tirado-Estrada et al. 2011; Bhasker et al. 2013a). However, the potential increase of enteric methane emission with adding enzyme observed in the present study does not support the opinion of mitigating enteric CH_4 using exogenous enzymes. Thus, more studies are needed to verify the mechanism of enzyme influences in methane production.

Plasma Parameters There was no difference among control, Treat 1, and Treat 2 in the concentration of ALT, GLU, LDH, TC, TG, TP, and UREA (Table 26). The concentration of α-AMY in Treat 2 tended to be lower than that of control and Treat 1.

Blood indicators are usually used as a tool to diagnose the health status of animals. In this study, supplementation of exogenous cellulase had no significant effect on dry matter intake and nutrient digestion, and no treatment effects on plasma ALT, GLU, LDH, TC, TG, TP, and α-AMY were observed, implying that the health status of cattle was similar for treatment and control. Kipriotis et al. (2015) also found no significant difference between enzyme treatment and control in serum concentration of *β*-hydroxybutyrate (BHB), total protein, total bilirubin, aspartate amino transferase, and gamma glutamyl transferase in dairy cows. While, Yang et al. (2000) and Lewis et al. (1996) found that the exogenous enzyme addition declined plasma concentration of BHB in the early and mid-lactation period, indicating a positive effect of the enzyme on body fat mobilization in this period of dairy cows. More studies are necessary to confirm the response of physical health status to exogenous fibrolytic enzyme addition.

3.3 Conclusion

In summary, exogenous cellulase and xylanase used in the current study did not influence feed intake, nutrient digestibility, energy utilization, and ruminal fermentation pattern in goats. Furthermore, the enzyme addition had no effect on the activities of ruminal extracellular and intracellular CMCase and xylanase, enteric methane emissions, and ruminal methanogens community diversity. Additionally, the present results indicated no synergism between exogenous cellulase and xylanase for improving of nutrient digestion, but xylanase application at a low dose increased the availability of retained and digested N in goats. The results clearly demonstrated that in vitro effects measured earlier are not confirmed under in vivo conditions. Moreover, it can be emphasized that application of exogenous enzyme products should focus on the application rate and mode, stability and source of enzymes, and enzyme feed specificity.

The inclusion of cellulase in the concentrate diet of cattle at a rate of 60 and 180 g/100 kg DM did not significantly enhance intake or digestion of DM, CP and fiber, rumen fermentation, methane emission, and blood parameters in the current trials. However, applying cellulase at a low-dose decreased the molar proportion of isobutyrate. The numerical increase in DM and CP digestion and methane emission imply that exogenous cellulase supplementation may have modulatory effects in improving nutrient utilization and increasing methane emission.

References

Arriola K, Kim SC, Staples CR, Adesogan AT (2011) Effect of fibrolytic enzyme application to low-and high-concentrate diets on the performance of lactating dairy cattle. J Dairy Sci 94:832–841

Beauchemin KA, Morgavi DP, McAllister TA, Yang WZ, Rode LM (2001) The use of enzymes in ruminant diets. In: Garnsworthy PC, Wiseman J (eds) Recent advances in animal nutrition. Nottingham University Press, Loughborough, pp 297–322

Beauchemin KA, Colombatto D, Morgavi DP (2004a) A rationale for the development of feed enzyme products for ruminants. Can J Anim Sci 84(1):23–36

Beauchemin KA, Colombatto D, Morgavi DP, Yang WZ, Rode LM (2004b) Mode of action of exogenous cell wall degrading enzymes for ruminants. Can J Anim Sci 84(1):13–22

Beauchemin KA, Kreuzer M, O̒Mara F, McAllister TA (2008) Nutritional management for enteric methane abatement: a review. Aust J Exp Agric 48(2):21–27

Beauchemin KA, McAllister TA, Mcginn SM (2009) Dietary mitigation of enteric methane from cattle. CAB Rev Perspect Agric Vet Sci Nutr Nat Res 4(35):1–18
Bhasker TV, Nagalakshmi D, Rao DS (2013a) Development of appropriate fibrolytic enzyme combination for maize stover and its effect on rumen fermentation in sheep. Asian Australas J Anim Sci 26:945–951
Bhasker TV, Nagalakshmi D, Srinivasa D, Raghunandhan T (2013b) Effect of supplementing exogenous fibrolytic enzyme cocktail on nutrient utilization in sheep fed on maize stover based total mixed ration. Indian J Anim Nutr 30:47–51
Bhat MK, Hazlewood GP (2001) Enzymology and other characteristics of cellulases and xylanases. In: Bedford M, Partridge G (eds) Enzymes in farm animal nutrition. CABI Publishing, Oxford, pp 11–60
Bowman GR, Beauchemin KA, Shelford JA (2002) The proportion of the diet to which fibrolytic enzymes are added affects nutrient digestion by lactating dairy cows. J Dairy Sci 85:3420–3429
Burroughs W, Woods W, Ewing SA, Greig J, Theurer B (1960) Enzyme additions to fattening cattle rations. J Anim Sci Technol 19:458–464
Chen J, Stokes MR, Wallace CR (1994) Effects of enzyme-inoculant systems on preservation and nutritive value of haycrop and corn silages. J Dairy Sci 77:501–512
Chen X, Mao HL, Wang JK, Wu CH, Liu JX (2013) Exogenous fibrolytic enzymes: effects on *in vitro* fermentation and methane production of corn silage. Chin J Anim Nutr 25(1):214–221
Chung YH, Zhou M, Holtshausen L, Alexander TW, McAllister TA, Guan LL, Oba M, Beauchemin KA (2012) A fibrolytic enzyme additive for lactating Holstein cow diets: ruminal fermentation, rumen microbial populations, and enteric methane emissions. J Dairy Sci 95(3):1419–1427
Colombatto D, Beauchemin KA (2003) A proposed methodology to standardize the determination of enzymic activities present in enzyme additives used in ruminant diets. Can J Anim Sci 83:559–568
Colombatto D, Morgavi DP, Furtado AF, Beauchemin KA (2003) Screening of exogenous enzymes for ruminant diets: relationship between biochemical characteristics and *in vitro* ruminal degradation. J Anim Sci Technol 81:2628–2638
Colombatto D, Mould FL, Bhat MK, Phipp RH, Owen E (2004) *In vitro* evaluation of fibrolytic enzymes as additives for maize (Zea mays L.) silage III. Comparison of enzyme derived from psychrophilic, mesophilic or thermophilic sources. Anim Feed Sci Technol 111:145–159
Dado RG, Allen MS (1995) Intake limitations, feeding behavior, and rumen function of cows challenged with rumen fill from dietary fiber or inert bulk. J Dairy Sci 78:118–133
Dong Y, Bae HD, McAllister TA, Mathison GW, Cheng KJ (1999) Effects of exogenous fibrolytic enzymes, α-bromoethanesulfonate and monensin on fermentation in a rumen simulation (RUSITEC) system. Can J Anim Sci 79(4):491–498
Eun JS, Beauchemin KA (2007) Enhancing *in vitro* degradation of alfalfa hay and corn silage using feed enzymes. J Dairy Sci 90:2839–2851
Eun JS, Beauchemin KA (2008) Assessment of the potential of feed enzyme additives to enhance utilization of corn silage fibre by ruminants. Can J Anim Sci 88(1):97–106
Eun JS, Beauchemin KA, Schulze H (2007) Use of exogenous fibrolytic enzymes to enhance *in vitro* fermentation of alfalfa hay and corn silage. J Dairy Sci 90:1440–1451
Gaafar HMA, Abdel-Raouf EM, EL-Reidy KFA (2010) Effect of fibrolytic enzyme supplementation and fiber content of total mixed ration on productive performance of lactating buffaloes. Slovak J Anim Sci 43:147–153
Giraldo LA, Tejido ML, Ranilla MJ, Carro MD (2007) Effects of exogenous cellulase supplementation on microbial growth and ruminal fermentation of a high-forage diet in RUSITEC fermenters. J Anim Sci 85(8):1962–1970
Giraldo LA, Tejido ML, Ranilla MJ, Carro MD (2008) Effects of exogenous fibrolytic enzymes on *in vitro* ruminal fermenation of substrates with different forage: concentrate ratios. Anim Feed Sci Technol 141:306–325
Goering HK, Van Soest PJ (1970) Forage fiber analysis (apparatus, reagents, procedures and some applications), Agriculture handbook no. 379. ARS-USDA, Washington, DC

He ZX, Yang LY, Yang WZ, Beauchemin KA, Tang SX, Huang JY, Zhou CS, Han XF, Wang M, Kang JH, Odongo NE, Tan ZL (2015) Efficacy of exogenous xylanases for improving *in vitro* fermentation of forages. J Agric Sci Cambridge 153(3):538–553

Hernández PA, Mendozaa GD, Bárcenab JR, Plataa FX, Martíneza JA, Leeb HA (2011) Effect of exogenous fibrolytic enzyme and slow release urea in finishing rations for lambs. J Appl Anim Res 39:104–107

Kipriotis E, Heping X, Vafeiadakis T, Kiprioti M, Alexopoulou E (2015) Ramie and kenaf as feed crops. Ind Crop Prod 68:126–130

Lewis GE, Hunt CW, Sanchez WK, Treacher R, Pritchard GT, Feng P (1996) Effect of direct-fed fibrolytic enzymes on the digestive characteristics of a forage-based diet fed to beef steers. J Anim Sci 74:3020–3028

Lewis GE, Sanchez WK, Hunt CW, Guy MA, Pritchard GT, Swanson BI, Treacher RJ (1999) Effect of direct-fed fibrolytic enzymes on the lactational performance of dairy cows. J Dairy Res 82:611–617

Lu Q, Jiao JZ, Tang SX, He ZX, Zhou CS, Han XF, Wang M, Kang JH, Odongo NE, Tan ZL (2015) Effects of dietary cellulase and xylanase addition on digestion, rumen fermentation and methane emission in growing goats. Arch Anim Nutr 69(4):251–266

Lu DX, Zhang PY (1996) Scientific technology of feeding goat. Beijing, China: China Agriculture Press; p. 71–126. Chinese

McAllister TA, Hristov AN, Beauchemin KA, Rode LM, Cheng KJ (2001) Enzymes in ruminant diets. In: Bedford MR, Partridge GG (eds) Enzymes in farm animal nutrition. CABI Publishing, Wallingford, pp 273–298

Morgavi DP, Beauchemin KA, Nsereko VL, Rode LM, McAllister TA, Iwassa AD, Wang Y, Yang WZ (2001) Resistance of feed enzymes to proteolytic inactivation by rumen microorganisms and gastrointestinal protease. J Anim Sci 79:1621–1630

Morgavi DP, Beauchemin KA, Nsereko VL, Rode LM, Mcallister TA, Wang Y (2004) Trichoderma enzymes promote fibrobacter succinogenes S85 adhesion to, and degradation of, complex substrates but not pure cellulose. J Sci Food Agric 84:1083–1090

Newbold CJ, Brock R, Wallace RJ (1992a) The effect of Aspergillus oryzae fermentation extract on the growth of fungi and ciliate protozoa in the rumen. Lett Appl Microbiol 15:109–112

Newbold CJ, Frumholtz PP, Wallace RJ (1992b) Influence of Aspergillus oryzae fermentation extract on rumen fermentation and blood constituents in sheep given diets of grass hay and barley. J Agric Sci Cambridge 119:423–427

Nsereko VL, Morgavi DP, Rode LM, Beauchemin KA, Callister TA (2000) Effects of fungal enzyme preparations on hydrolysis and subsequent degradation of alfalfa hay fiber by mixed rumen microorganisms *in vitro*. Anim Feed Sci Technol 88:153–170

Nsereko VL, Beauchemin KA, Morgavi DP, Rode LM, Furtado AF, Mcallister TA, Iwaasa AD, Yang WZ, Wang Y (2002) Effect of a fibrolytic enzyme preparation from Trichoderma longibrachiatum on the rumen microbial population of dairy cows. Can J Microbiol 48:14–20

Oba M, Allen MS (1999) Evaluation of the importance of the digestibility of neutral detergent fiber from forage: effects on dry matter intake and milk yield of dairy cows. J Dairy Sci 82:589–596

Oba M, Allen MS (2000) Effects of brown midrib 3 mutation in corn silage on productivity of dairy cows fed two concentrations of dietary neutral detergent fiber: 3. Digestibility and microbial efficiency. J Dairy Sci 83:1350–1358

Peters A, Lebzien P, Meyer U, Borchert U, Bulang M, Flachowsky G (2010) Effect of exogenous fibrolytic enzymes on ruminal fermentation and nutrient digestion in dairy cows. Arch Anim Nutr 64:221–237

Phipps RH, Sutton JD, Beever DE, Bhat MK, Hartnell GF, Vicini JL, Hard DL (2000) Effect of cell-wall degrading enzymes and method of application on feed intake and milk production of Holstein-Friesian dairy cows. J Dairy Sci 83(Suppl. 1):23. (Abstr.)

Reddish MA, Kung LJ (2007) The effect of feeding a dry enzyme mixture with fibrolytic activity on the performance of lactating cows and digestibility of a diet for sheep. J Dairy Sci 90:4724–4729

Rode LM, Yang WZ, Beauchemin KA (1999) Fibrolytic enzyme supplements for dairy cows in early lactation. J Dairy Sci 82:2121–2126

Russi JP, Wallace RJ, Newbold CJ (2002) Influence of the pattern of peptide supply on microbial activity in the rumen simulating fermenter (RUSITEC). Br J Nutr 88:73–80

Salter DN, Daneshvar K, Smith RH (1979) The origin of nitrogen incorporated into compounds in the rumen bacteria of steers given protein- and urea- containing diets. Br J Nutr 41(1):197–209

Satter LD, Slyter LL (1974) Effect of ammonia concentration on rumen microbial protein production *in vitro*. Br J Nutr 32(2):199–208

Shekhar C, Thakur SS, Shelke SK (2010) Effect of exogenous fibrolytic enzymes supplementation on milk production and nutrient utilization in Murrah buffaloes. Trop Anim Health Prod 42:1465–1470

Stewart CS, Flint HJ, Bryant MP (1997) The rumen bacteria. In: Hobson PN, Stewart CS (eds) The rumen microbial ecosystem. Chapman and Hall, London, pp 10–72

Sun ZH, Liu SM, Tayo GO, Tang SX, Tan ZL, Lin B, He ZX, Hang XF, Zhou ZS, Wang M (2009) Effects of cellulase or lactic acid bacteria on silage fermentation and *in vitro* gas production of several morphological fractions of maize stover. Anim Feed Sci Technol 152:219–231

Sutton JD, Phipps RH, Beever DE, Humphries DJ, Hartnell GF, Vicini JL, Hard DL (2003) Effect of method of application of a fibrolytic enzyme production digestive processes and milk production in Holstein-Friesian cows. J Dairy Sci 86:546–556

Tang S, Tan Z, Zhou C, Jiang H, Jiang Y, Sheng L (2006) A comparison of *in vitro* fermentation characteristics of different botanical fractions of mature maize stover. J Anim Feed Sci 15(3):505–515

Tang SX, Tayo GO, Tan ZL, Sun ZH, Shen LX, Zhou CS, Xiao WJ, Ren GP, Han XF, Shen SB (2008) Effects of yeast culture and fibrolytic enzyme supplementation on *in vitro* fermentation characteristics of low-quality cereal straws. J Anim Sci Technol 86:1164–1172

Tang SX, Zou Y, Wang M, Salem AZM, Odongo NE, Zhou CS, Han XF, Tan ZL, Zhang M, Fu YF, Huang SQ, He ZX, Kang JH (2013) Effects of exogenous cellulase source on *in vitro* fermentation characteristics and methane production of crop residues and grasses. Anim Nutr Feed Technol 13:111–126

Tenkanen M, Schuseil J, Puls J, Poutanen K (1991) Production, purification and characterization of an esterase liberating phenolic acids fromlignocellulosics. J Biotechnol 18:69–83

Tirado-Estrada G, Mendoza-Martínez GD, Pinos-Rodríguez JM, Quezada-Tristán T, Guevara-Lara F (2011) Effects of two fibrolytic enzyme mixtures on growth performance, digestion and ruminal fermentation in lambs fed corn stover based diets. J Appl Anim Res 39:158–160

Wang M, Tang SX, Tan ZL (2011) Modeling *in vitro* gas production kinetics: derivation of logistic-exponential (LE) equations and comparison of models. Anim Feed Sci Technol 165(3–4):137–150

Wang M, Sun XZ, Tang SX, Tan ZL, Pacheco D (2013) Deriving fractional rate of degradation of logisticexponential (LE) model to evaluate early *in vitro* fermentation. Animal 7:920–929

Yang WZ, Beauchemin KA, Rode LM (1999) Effects of enzyme feed additives on extent of digestion and milk production of lactating dairy cows. J Dairy Sci 82:391–403

Yang WZ, Beauchemin KA, Rode LM (2000) A comparison of methods of adding fibrolytic enzymes to lactating cow diets. J Dairy Sci 83:2512–2520

Zhou M, Chung YH, Beauchemin KA, Holtshausen L, Oba M, McAllister TA, Guan LL (2011) Relationship between rumen methanogens and methane production in dairy cows fed diets supplemented with a feed enzyme additive. J Appl Microbiol 111(5):1148–1158

Zhu Y, Nishino N, Kishida Y, Uchida S (1999) Ensiling characteristics and ruminal degradation of Italian ryegrass and Lucerne silages treated with cell wall-degrading enzymes. J Sci Food Agric 79:1987–1992

MIX
Papier aus verantwortungsvollen Quellen
Paper from responsible sources
FSC® C105338

If you have any concerns about our products,
you can contact us on
ProductSafety@springernature.com

In case Publisher is established outside the EU,
the EU authorized representative is:
Springer Nature Customer Service Center GmbH
Europaplatz 3, 69115 Heidelberg, Germany

Printed by Libri Plureos GmbH
in Hamburg, Germany